TRAVEL IN TOWNS

TRAVEL IN TOWNS

Jam yesterday, jam today and jam tomorrow?

MARTIN J. H. MOGRIDGE

M
MACMILLAN
REFERENCE
BOOKS

First published 1990 by
THE MACMILLAN PRESS LTD
London and Basingstoke

Associated companies in Auckland, Delhi, Dublin, Gaborone, Hamburg, Harare, Hong Kong, Johannesburg, Kuala Lumpur, Lagos, Manzini, Melbourne, Mexico City, Nairobi, New York, Singapore, Tokyo.

British Library Cataloguing in Publication Data

Mogridge, Martin
Travel in towns: jam yesterday, jam today and jam tomorrow?
1. Great Britain, Road traffic, Planning
I. Title
711. 730941

ISBN 0-333-53204-X

Printed in Great Britain by
Billing & Sons Ltd, Worcester

Contents

List of Tables

List of Figures

In the long run the most potent factor in maintaining a 'ceiling' on private car traffic in busy areas is likely to be the provision of good, cheap public transport, coupled with the public's understanding of the position. This last is essential, and is indeed one of the reasons why we have endeavoured to write this Report in non-technical language for the general reader. But the attractions of the private car are very great, and there can be no denying the difficulties of providing public transport services so intrinsically *convenient* that they will attract optional car traffic off the roads in appreciable quantities. But, given a different financial policy, travel by public transport could be made *relatively cheap*, and this may prove to be the key to the problem in the long term.

The Buchanan Report
'Traffic in Towns'
1963

'The rule is, jam to-morrow and jam yesterday—but never jam to-day.'

'It *must* come sometimes to "jam to-day",' Alice objected.

'No, it can't,' said the (White) Queen. 'It's jam every *other* day: to-day isn't any *other* day, you know.'

Lewis Carroll
'Through the Looking Glass
and What Alice Found There'

'Jam yesterday, jam today, but please no jam tomorrow'

Pathé News
'The Great Hold-up'
1953

Foreword

This book has been written as a summary of a series of research papers produced over the last six years on how to improve travel in towns. Since the publication of my lecture 'Jam yesterday, jam today and jam tomorrow?', a University College London Lunch-Hour Lecture given on October 17th 1985, the controversy over my re-examination of some very old ideas about travel in towns has caused a number of researchers to look again at the fundamental principles underlying the way transport planning is and has been done in the U.K. over the last 30 years. This book is, therefore, a restatement of my position in the light of the debate from my colleagues in the profession.

Moreover, the transport problem in London, in particular, has become of increasing concern over the last few years. This is mainly because the number of employees coming in to the centre to work every day has grown rapidly since 1982, reversing a 20-year decline of a quarter in the space of seven years. This has given a new focus to the debate about the place of public transport in the total transport problem of a city like London and to the methods of solving the problems of traffic congestion, both on the roads and on the public transport systems.

As a symptom of this renewed interest, there has been a great debate in the media, with many documentaries on both T.V. and radio, and features in the press, about various aspects of the transport problem. I have chosen, therefore, to comment in this book on some of the popular misconceptions about transport policy—and to try to explain some of the complexities of behaviour—which have led many people into fallacious interpretations of events.

I would like to thank the various people and organisations who have given me financial support during the development of these ideas. As a contract researcher and not a permanent staff member of the university, I rely totally on such support. It is encouraging that such support is forthcoming, even when the ideas are as controversial, though of long standing, as the ones I am expressing in this book.

The organisations include the Transport and Road Research Laboratory, British Rail, London Regional Transport, Greater London Council, Docklands Forum, Transport 2000, Transnet, Rees Jeffreys Road Fund Trust and the Economic and Social Research Council.

Prologue

Every technology goes through three stages: first, a crudely simple and quite unsatisfactory gadget; second, an enormously complicated group of gadgets designed to overcome the shortcomings of the original, and achieving thereby somewhat satisfactory performance through extremely complex compromise; third, a final stage of smooth simplicity and efficient performance based on correct understanding of natural laws and proper design therefrom.

In transportation, the oxcart and rowboat represent the first stage of technology.

The second stage might well be represented by the automobiles of the middle twentieth century just before the opening of interplanetary travel. These unbelievable museum pieces were for the time fast, sleek and powerful—but inside their skins was assembled a preposterous collection of mechanical buffoonery. The prime mover for such a juggernaut might have rested in one's lap; the rest of the mad assembly consisted of afterthoughts intended to correct the uncorrectable, to repair the original basic mistake in design—for automobiles and even the early aeroplanes were 'powered' (if one may call it that) by 'reciprocating engines'.

A reciprocating engine was a collection of miniature heat engines using (in a basically inefficient cycle) a small percentage of an exothermic chemical reaction, a reaction which was started and stopped every split second. Much of the heat was intentionally thrown away into a 'water jacket' or 'cooling system', then wasted into the atmosphere through a heat exchanger.

What little was left caused blocks of metal to thump foolishly back-and-forth (hence the name 'reciprocating') and thence through a linkage to cause a shaft and flywheel to spin around. The flywheel (believe it if you can) had no gyroscopic function; it was used to store kinetic energy in a futile attempt to cover up the sins of reciprocation. The shaft at long last caused wheels to turn and thereby propelled this pile of junk over the countryside.

The prime mover was used only to accelerate and to overcome 'friction'—a concept then in much wider engineering use. To decelerate, stop, or turn the heroic human operator used his own muscle power, multiplied precariously through a series of levers.

Despite the name 'automobile', these vehicles had no autocontrol circuits; control, such as it was, was exercised second by second for hours on end by a human being peering out through a small pane of dirty silica glass, and judging unassisted and often disastrously his own motion and those of other objects. In almost all cases the operator had no notion of the kinetic energy stored in his missile and could not have written the basic equation. Newton's Laws of Motion were to him mysteries as profound as the meaning of the universe.

Nevertheless millions of these mechanical jokes swarmed over our home planet, dodging each other by inches or failing to dodge. None of them ever worked right; by their nature they could not work right; and they were constantly getting out of order. Their operators were usually mightily pleased when they worked at all. When they did not, which was every few hundred miles (hundred, not hundred thousand), they hired a member of a social class of arcane specialists to make inadequate and always expensive repairs.

Despite their mad shortcomings, these 'automobiles' were the most characteristic form of wealth and the most cherished possessions of their time. Three whole generations were slaves to them.

Robert Heinlein
Space Family Stone
1952

1 Introduction

Traffic speeds

In March 1983, I became involved in the up-dating of the survey carried out periodically by the Transport Studies Group at University College on traffic speeds in central London. This survey was begun in 1962, and comprises a tour around a set of 25 random points in a 10-kilometre square centred on a centre of London (the national grid centre). The MSc students of the Group travel around this tour a number of times, both by car and by bus.

Such a random tour does not represent actual travel, but rather enables an expression of potential travel quality to be derived. Repeated surveys thus measure whether travel quality is deteriorating or not through time.

Travellers, however, are more concerned with their actual journeys. Given the spatial structure of the city, high flows of travellers are to be found in the central area where employment and activities are most concentrated. A survey of the traffic speeds actually experienced will take into account the traffic flows along the links measured, in order to weight correctly the average traffic speed.

Even this is not strictly sufficient, because the traffic flows will be counts of vehicles and not counts of people travelling. Nevertheless, the assumption that the occupancy rate for a car does not change over the area and time period being surveyed is usually adequate. That for a bus, on the other hand, may well change substantially as the bus passes through the areas of intense concentration of activities.

What I found remarkable, as I became more engrossed in the data on traffic speeds, was that traffic speeds had remained approximately constant in London over a long period of time during which car ownership had risen substantially, and substantially faster than road capacity had increased.

The 'barely tolerable' speed

This fact seemed to require a better explanation than those then currently accepted as valid, namely that traffic speeds settled down to a 'barely tolerable' or 'acceptable' level in the centres of cities.

In particular, there seemed to be no explanation in this concept of why traffic speeds rose at a certain rate as one progressed outwards from the centre of the city, apart from the supposition that people, sometimes the same people, preferred to drive at different speeds at different distances from the city centre.

I thought this explanation palpably absurd and, thus stimulated, began my search for a better explanation.

The growth in car ownership

My search ran in parallel with my concerns about the consequences of the continued expansion of car ownership for the use of the space of the city. My first forecast of car ownership was published in 1967, as an invited paper for the first issue of the *Journal of Transport Economics and Policy* (*JTEP*).

I showed in that paper how the expenditure patterns of households led to the conclusion that car ownership would rise to much higher levels than then current government forecasts. Since then, the oil crisis of 1973 led to a faltering of the growth, and a controversy about forecasting methods, which I covered in my book on *The Car Market* published in 1983.

Government forecasts of car ownership and traffic since then have continually been revised upwards, culminating in the 1989 forecasts of between 83 and 142 percent increases in traffic above the 1988 value by 2025 (Department of Transport, 1989a).

Curbing traffic growth

In my paper in 1989 for the 21st anniversary commemorative issue of the *JTEP*, updating my original work, I argued that there were now two features of the way people travel which could be used to counter the growth in car traffic. Both are basically that a better alternative to car travel could be offered to potential travellers.

Car travel is essentially a short-distance, slow method of travel, with national average speeds of 21 mph and average distances of 7 miles, with the mode at no more than 3 miles. In London, average speeds are only slightly slower at 18 mph, with an average distance travelled of 6 miles.

One feature which could be used was that only 4 percent of car travel is on journeys over 25 miles long, although this accounts for over 33 percent of car distance travelled, but it is travelled at an average speed of no more than 33 mph. The creation of a high-speed rail network in the U.K., and indeed in Europe, would, I argued, set an upper limit to the level of motorway use by offering a better alternative to such longer distance travel.

The other feature was the amount of car travel within, and to and from, the centres of cities. Although the quantity of car travel in the suburbs of our cities is growing rapidly, the spatial growth of the cities themselves means that more spatially efficient methods of travel are required within, and to and from, the centres, which means better public transport.

The use of space

Cars are very inefficient at using the space of the city, requiring much more space—not only for travel but even more so for parking—than the space required for the activity itself which is the reason for the travel.

This was clear to some planners in the early stages of the growth of car ownership, culminating in the extremely influential report by Buchanan in 1963 *Traffic in Towns*. Buchanan, as had some traffic engineers before him such as Smeed, had shown quite clearly that it was physically impossible to accommodate everyone in the city centre if they were all to travel by car, even with total rebuilding of the centre.

Nevertheless, other planners, such as those who prepared the Greater London Development Plan (GLDP) of 1969, were convinced that a massive road-building programme was required to improve the facilities for car-borne travellers. A series of rings and radials were designed, much in the style of those that had been designed in a number of earlier plans for London from 1906 onwards, culminating in Abercrombie's Greater London Plan of 1944.

When I joined the Greater London Council (GLC) in 1966, at the formation of its transport research team, one of my first jobs was to implement on the GLC's computers the analysis package used to determine the traffic flows on the proposed motorway network.

At that time, the consultants used in the 1962 London Transportation Study were still in the process of predicting future traffic demands. It was very evident that the motorway network, at the design speeds, could not possibly cope with the volume of traffic that would want to use them, and various arbitrary mathematical procedures were used to limit the amount of forecast traffic to what the motorways could indeed carry.

Those who were convinced that a motorway network was necessary could point to the fact that they would be running full and say: 'Look! Of course we need them. Look how full they are!'

At the same time, the arbitrary procedures used to limit forecast car traffic meant that these potential travellers had to travel some other way and therefore that improvements to public transport had to be made. This was never satisfactorily resolved during the Inquiry into the GLDP, nor indeed by the time it was approved by the Government in 1976.

'Homes before Roads'

By 1976, however, the motorway proposals had been rejected by the GLC electorate in the 1973 election, following a campaign against them which used as its foundation the work of Thomson in *Motorways in London* of 1969. This had pointed to the extensive improvements to the secondary roads that would have to be made as well as the construction of the primary motorway routes themselves, entailing substantial demolition of property. Thus the 'Homes before Roads' campaign motif.

A second major study, the 'London Rail Study' under Barran, had reported in 1974 on possible changes to the rail system, but, like its predecessor in 1968, had concluded there was no case for substantial rail investments.

Thus the improvements to public transport were not forthcoming, and, indeed, with the financial crisis of the middle 1970s, emphasis was placed on keeping fares down and increasing subsidy levels rather than on investment to make the system more viable. I was, during the period 1973–78, the Principal Planner leading the Economics Section of the GLC's Transportation Branch and thus was involved in assessing the relative values of fares subsidy against capital investment.

I felt, at this time, that there was something basically wrong with the process used in the transportation model to assess future traffic demands, but was unable to put my finger on it. Some of my colleagues (Goodwin, Martin and Talbot, 1977) produced an exhaustive analysis of the modelling process, and its application in London, but this still seemed to me to have a key part missing.

The missing key

It was not until 1983—when I could ask the question 'why are traffic speeds so stable?'—that I began to see what the missing key was.

This book is the result of my search for a better explanation to that question and what the key to a better transportation planning process is. I found it essentially in the work of Pigou in 1920 and Knight in 1924.

Pigou argued as follows (p.194):

> 'Suppose there are two roads ABD and ACD both leading from A to D. If left to itself, traffic would be so distributed that the trouble involved in driving a "representative" cart along each of the two roads would be equal. But, in some circumstances, it would be possible, by shifting a few carts from route B to route C, greatly to lessen the trouble of driving those still left on B, while only slightly increasing the trouble of driving along C. In these circumstances a rightly chosen measure of differential taxation against road B would create an "artificial" situation superior to the "natural" one. But the measure of differentiation must be rightly chosen'.

Knight amplified these points in the following way (pp.584-85):

> 'In Professor Pigou's study the argument that free enterprise leads to excessive investment in industries having relatively upward-sloping cost curves is developed with the aid of a concrete example, the case of two roads.
>
> 'Suppose that between two points there are two highways, one of which is broad enough to accommodate without crowding all the traffic which may care to use it, but is poorly graded and surfaced, while the other is a much better quality road but narrow and quite limited in capacity. If a large number of trucks operate between the two termini and are free to choose either of the two routes, they will tend to distribute themselves between the roads in such proportions that the cost per unit of transportation, or effective result per unit of investment, will be the same for every truck on both routes.
>
> 'As more trucks use the narrower and better road, congestion develops, until at a certain point it becomes equally profitable to use the broader but poorer highway. The congestion and interference resulting from the addition of any particular truck to the stream of traffic on the narrow but good road affects in the same way the cost and output of all the trucks using that road.
>
> 'It is evident that if, after equilibrium is established, a few trucks should be arbitrarily transferred to the broad road, the reduction in cost, or increase in output, to those remaining on the narrow road would be a clear gain to the traffic as a whole. The trucks so transferred would incur no loss, for any one of them on the narrow road is a marginal truck, subject to the same relation between cost and output as any truck using the broad road. Yet whenever there is a difference in the cost, to an additional truck, of using the two roads, the driver of any truck has an incentive to use the narrow road, until the advantage is reduced to zero for all the trucks.
>
> 'Thus, as the author contends, individual freedom results in a bad distribution of investment between industries of constant and industries of increasing cost.
>
> 'In such a case social interference seems to be clearly justified. If the government should levy a small tax on each

truck using the narrow road, the tax would be considered by the trucker as an element in his cost, and would cause the number of trucks on the narrow road to be reduced to the point where the *ordinary cost, plus the tax*, became equal to the cost on the broad road, assumed to be left tax free.

'The tax could be so adjusted that the number of trucks on the narrow road would be such as to secure the maximum efficiency in the use of the two roads taken together. The revenue obtained from such a tax would be a clear gain to the society, since no individual truck would incur higher costs than if no tax had been levied'.

A diagram of costs against flow, in Figure 1.1, on the two roads may help to make this point in another way. As flow increases from zero, the narrow road gradually fills up, with costs increasing faster and faster with each additional truck, until costs on the two roads are equal. All further traffic then uses the broad road.

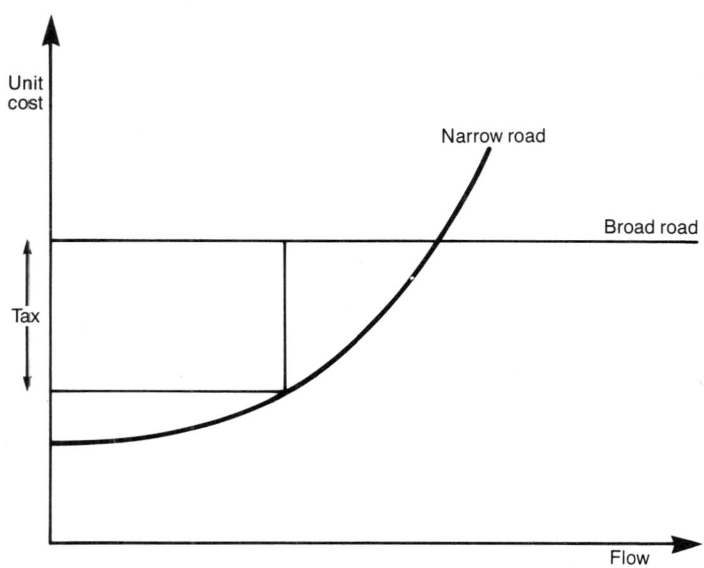

Figure 1.1 Cost/flow relationships on two competing highways

Beyond this demand level, therefore, the costs on the narrow road are totally dependent on the fixed costs of the broad road, and not at all on the fixed costs of the narrow road itself.

Only if demand is insufficient to fill the narrow road up to the level at which some trucks use the broad road will costs on the narrow road be independent of the costs on the broad road.

Cars versus public transport

Now, while these two arguments are put in terms of carts and trucks respectively, they are immediately applicable to our problem of moving people in a congested city centre if we consider one route to be car travel and the other to be public transport of one kind or another, with the proviso that both refer to complete door-to-door journeys.

Car travel has lower fixed costs, because, when there is no one else on the road, car travel is quintessentially a point-to-point on-demand non-stop method of travel.

It is intriguing to note that this concept of 'no one else on the road' is intrinsic to the current argument about the costs of traffic congestion, since the cost of congestion is taken to be the difference between the fixed costs when no one else is on the road and the actual level of cost at the actual traffic level.

Public transport, on the other hand, has higher fixed costs, because the service is quintessentially a stop-to-stop or station-to-station method of travel, requiring additional walking at each end, stopping at intermediate stops or stations, providing a service at fixed frequencies and with a fixed network spacing, and, moreover, often requiring interchange between services.

Public transport is therefore usually a slower method of travel than the car and thus has higher time costs than the car (time multiplied by the value of time to the traveller) when there is no one else on the road or service, except for certain non-stop, longer distance services by rail.

The fixed costs of public transport also generally include a share of the overheads of running the service, including the costs of the staff and the depreciation of assets, unlike car travel where the costs of ownership are often not included in the assessment of the costs of a journey by the travellers themselves.

Nevertheless, the basic difference between car travel and public transport is that the former has high variable costs (i.e. costs rise ever more steeply as congestion develops), whereas the latter has essentially low variable costs; that is, costs per passenger are independent of flow levels.

This corresponds to the situation of a narrow but limited capacity road and a broad, high-capacity but slow road respectively in the argument of Pigou and Knight above.

Thus, if there are enough car drivers, the costs on the roads are driven up to the level of the costs on the public transport system, and no road-building programme can reduce the car costs below the level of public transport costs unless it is substantial enough to take all the offered demand, which as we mentioned above is physically impossible.

The situation for city centres is actually worse than this. Public transport does, in fact, not have a zero variable cost but a negative variable cost; that is, the cost of a journey to an individual traveller goes down as the flow increases. This is because a higher demand level leads the public transport operator to increase his service frequency and reduce his network spacing (run more routes). The operator is also able to spread his overhead costs over a bigger number of passengers.

This is a general consequence of a scale economy; that is, costs per unit reduce if the scale of operations increases. It was examined in detail for buses by Mohring in 1972.

Road pricing

Knight argued that a judicious choice of taxation imposed on the users of the high variable cost route would reduce the consumption of resources without making anyone worse off (since the remaining users would be paying no more in total than before), but would obtain a higher speed of travel for an additional price.

This aspect of the equilibrium is often ignored in the road pricing argument, but becomes especially relevant when one is considering the equilibrium between public and private users of the roads rather than the railways.

Edgeworth in 1925, extended by Hotelling in 1932, took the argument further into this area by demonstrating that there was a tax level which could be levied on one of two alternative commodities which would lower the price of both. In its application to the road pricing problem, a tax levied on a road with an upward-sloping cost curve, or its inverse, a subsidy paid to public transport with a downward-sloping cost curve, would lower the costs of travel on both.

The key point here is that cars are, as already noted, inefficient users of space compared to communal road vehicles. There is therefore a taxation level which would optimize the use of the space of the road system and leave the remaining car drivers better off than before; public transport passengers, both those remaining and those who had shifted from car, would also be better off than before.

This is a complex issue because there is a difference between cars and public transport in the sense that the high-capacity forms of the latter (trams, buses and minibuses) have to operate on segregated routes for maximum efficiency.

If they are to operate in this way, and not just in incomplete bus-lanes, then the road-pricing and public transport subsidy scheme has to be coupled with the road-allocation scheme to give maximum efficiency.

I came into the argument, however, by realising the force of the other extension to Knight's argument, that road-building in city centres, by attracting people off the public transport system, can increase the public transport costs and thus, by the equilibrium process, increase the costs of the road system.

The Downs-Thomson paradox

This seems to have been first pointed out by Downs in 1962, followed by Thomson in 1977, with Suchorzewski in 1973 giving the converse; that is, to reduce road costs you have to improve public transport. We named this effect the Downs-Thomson paradox in a paper I wrote with my colleagues in 1987.

This argument that there is an equilibrium between car and public transport costs seemed to me to give a better explanation of the stability of traffic speeds in the centre, and to and from the centre, of cities like London. Most of the first part of this book is an examination of the detail of the argument to verify whether or not the proposition can be valid.

The consequences of the argument for the evaluation process of the standard transport model are also examined in detail. Rather than by using arbitrary mathematical processes to reduce demand, or to shift demand from one mode to another, there is a perfectly logical and self-consistent way in which this can and should be done.

System optimisation

If the individual choices of the travellers choosing the minimum cost route and mode for themselves, aggregated over all travellers, are to give the minimum cost for the system as a whole, and thus the maximum efficiency of the transport system, then the intervention of a directing intelligence is required to allocate road space and to decide on the level of road price and public transport subsidy.

The standard transport planning modelling process recognizes that, indeed is based on the fact that, people choose the transport that has the minimum cost for themselves. There was, however, no process whereby the directing intelligence could work towards the system minimum cost, merely a hit and miss process of trying various options for improvement to see which gave the best result.

This is mainly because transport modelling was initially seen in terms of choosing investment strategies, with benefit/cost analysis the evaluation technique for choosing the best investment. It has been only gradually that pricing policies have been incorporated into the evaluation process in a comprehensive manner.

There is, however, a perfectly logical method of optimising the levels of taxation and subsidy. This requires the levels to be set at the level which forces people to face the costs of consumption of the resources actually used in making the journey, in other words the marginal cost of their journey. Where the marginal cost is greater than the perceived cost, as in car journeys in city centres, then a tax must be levied equal to the difference between the costs. Where the marginal cost is less than the perceived cost, as in public transport journeys, then a subsidy must be paid equal to the difference between the costs. The transportation planning modelling process must therefore work in terms of marginal costs, not perceived costs, if it is to reach a system optimum.

Most of the second half of this book is about the development of transportation modelling in London as these various ideas have been gradually incorporated into the modelling process.

I hope that, as a result of the work outlined in this book, it will become much clearer what such a system minimum consists of in the case of a congested city like London, and that the ideas of the economists named above will finally be applied to improve the transport system so that it operates at maximum efficiency.

System dynamics

It cannot be, however, a static process. There is no such thing as an end-state, a perfect transport system. Developments in automation and control, and in new concepts of moving people, and in motive power, will continually change the relative costs of the various modes of transport.

Moreover, the city will itself change in response to the changes in transport provided, leading to new demands for space and movement. London, like other cities, has been expanding outwards ever since the transport revolutions began with the introduction of horse-drawn stage coaches. With the introduction of high-speed trains, the commuter area is now rapidly spreading over South-East England and beyond, even, in the near future, to North-West France as the Channel Tunnel is completed.

Since the building of new transport networks is an expensive and time consuming process, we need to forecast as far ahead as we can reasonably see, in order to anticipate these new demands, and to channel them into the most efficient forms for serving the city.

Transport revolutions

I attempt some speculations on this theme towards the end of the book. Essentially, I accept the view that transport revolutions seem to occur in waves every fifty or so years, the Kondratiev cycle theory.

The revolution brought by the steam engine and railways transformed the London of the 1840s and 1850s; that of the electric engine and underground tubes transformed central London between 1890 and 1908; that of the electric tram, and the petrol engine and motor buses and cabs, rapidly replaced horse-drawn vehicles in inner London between 1901 and 1914; that of the car saw a quadrupling of ownership between 1949 and 1964.

All these transport revolutions, therefore, began to have their major impacts within a decade or so, although the ramifications continued for many years afterwards.

I believe that we are on the threshhold of another major change in the transport system of London and other cities.

The electronic revolution will change the car to be a completely automatic vehicle, a true automobile, able to take its passengers by itself wherever they want to go in complete safety merely by voice control. This opens up the car to every mobile person, not merely to those who can drive. It changes fundamentally our relationship with the car, from an artefact with connotations of status and power to a mere artefact that can be used or not, by anyone, if appropriate for a particular journey.

Such a revolution does not, however, change the fundamental problem of city transport, namely that the car is an inefficient user of space. The space revolution, which is the main concern of this book, will, I believe, lead to a fundamental reappraisal of the use of the space of the city in the next decade or so, and the change in the use of space to improve the efficiency of public transport, and thus of the transport system as a whole.

Such a change requires a directing intelligence at communal level and cannot be achieved by individuals acting on their own. As Buchanan so rightly pointed out in 1963:

> 'In the long run the most potent factor in maintaining a "ceiling" on private car traffic in busy areas is likely to be the provision of good, cheap public transport, coupled with the public's understanding of the position. This last is essential, and is indeed one of the reasons why we have endeavoured to write this Report in non-technical language for the general reader. But the attractions of the private car are very great, and there can be no denying the difficulties of providing public transport services so intrinsically *convenient* that they will attract optional car traffic off the roads in appreciable

> quantities. But, given a different financial policy, travel by public transport could be made *relatively cheap*, and this may prove to be the key to the problem in the long term'.

Whilst my emphasis is, as will be appreciated, different to that of Buchanan, his sentiments pre-echo mine precisely.

2 *Road Speeds*

In March 1983, I became involved in the latest in a series of surveys which had begun in 1962 to measure the travel speeds in central London. This was carried out by the MSc students and others of the Transport Studies Group of University College London, of which I was a member. There has in fact been a subsequent survey in 1988.

In 1962, Holroyd and Scraggs had defined a set of 25 random points within a 10-km square centred on the national grid centre at Charing Cross (the centre is at 530180; actually, it is in Horse Guards Parade and only metres away from 10 Downing Street). A tour round these 25 points, in the order in which they were generated by the random number generator, then enables the travel speed to be determined. The points are actually the nearest road junction to the mid-point of the 100-metre square defined by the random number. Hectares without roads, such as open or derelict spaces, are passed over. The same 25 points are used in each subsequent survey.

The tours are actually done by both car and bus, during the off-peak (i.e. between 9.30 am and 4.30 pm); my contribution in 1983 was to pilot a tour by rail for the first time. It took me three days of jumping in and out of trains every few minutes to go round the tour once. We will see why I did this in the next chapter.

A map of the 10-km square is shown in Figure 2.1 with the random points marked. The tour round these points is shown in Figure 2.2.

It can be seen that not only can this tour be used to define on-the-road speeds, it can also be used to define the direct speeds; that is, the speeds of travel in a straight line. After all, travellers want to get between an origin and a destination to do something; it is not often that they want merely to travel.

More specifically, this random tour gives us a measure of the quality of the road system for potential travellers, rather than the actual travellers, and as such can give us, in principle, a better

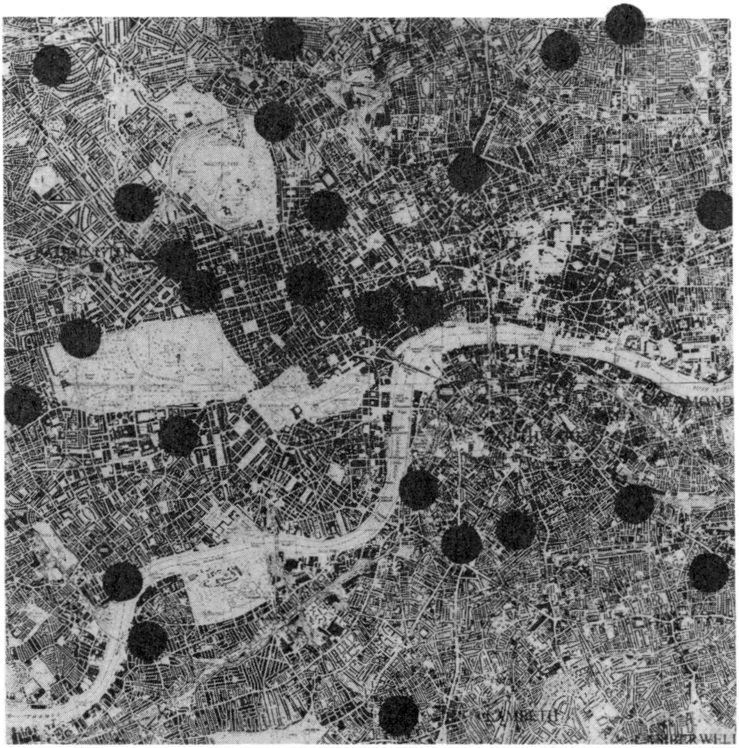

Figure 2.1 The random points in the 10-km square

representation of the changes through time, either enhancement or deterioration.

In fact, however, because of the great variability in day-to-day travel speeds on a given route, the number of runs needed to give a good representation is actually quite high, and much higher than the number done in our survey. In 1983, between five and 10 runs were done on each section of the tour (Ferguson, 1985). Formerly, only one run was done in 1962, and only two in 1969, 1972 and 1977, the previous surveys (Holroyd and Scraggs (1964), Yates and Howerd (1969), Buckles (1973), Nicholl (1977)). In 1988, between three and five runs were done (Heydecker and Ferguson, 1990).

The car journey speeds are obtained by the 'floating car' technique; that is, the driver has to overtake as many cars as overtake him or her to maintain his or her place in the traffic stream. They

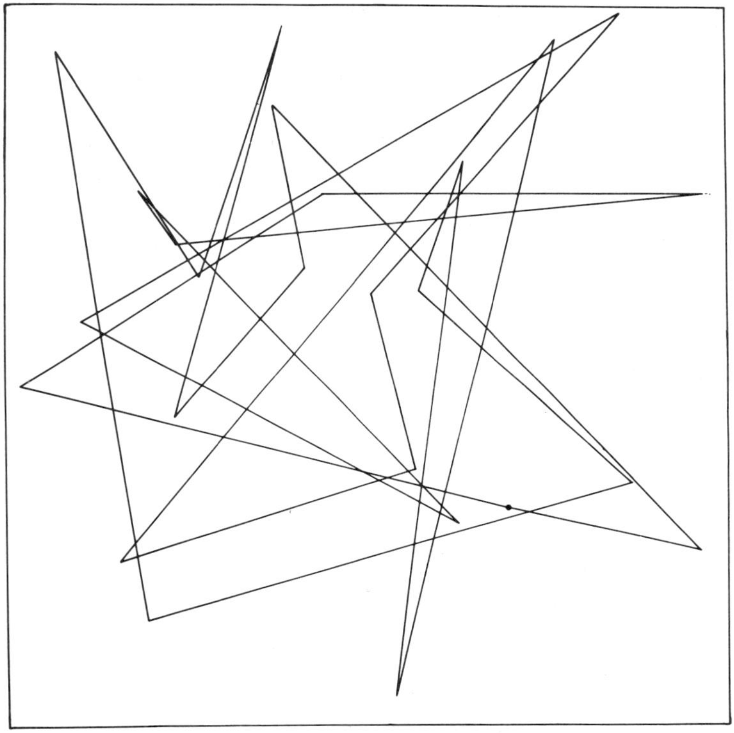

Figure 2.2 The random tour

are, in fact, what is known as kerb-to-kerb speeds. As well as the major routes of the central area, they might include at each end of the journey a small part which might be on a residential or local road.

It can be seen in the following table that the random journey speeds by car, on this kerb-to-kerb basis, have been recorded as varying up and down over time. This is probably spurious as we shall go on to show presently.

The route factor appears to have increased between 1969 and 1972, after the extensive introduction of one-way systems in the 1960s.

A plot of the route factor against the direct distance for the 1963 survey, given in Figure 2.3 below, shows that there is considerable

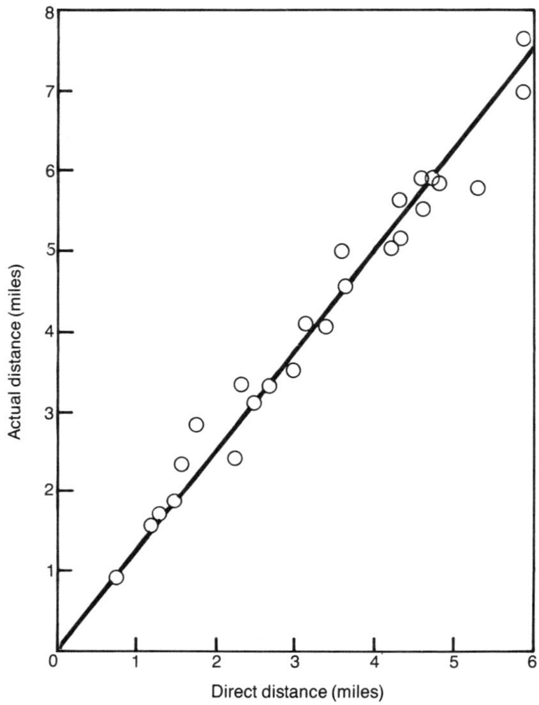

Figure 2.3 The route factor for the random journeys in the 10-km square

Table 2.1 Central London random journey speeds (miles/h)

	1963	*1969*	*1972*	*1977*	*1983*	*1988*
Journey speed	12.7	13.2	15.1	13.0	11.8	11.1
Direct speed	10.2	10.7	11.2	9.6	9.2	8.3
Route factor	1.25	1.23	1.34	1.36	1.28	1.34

variation for individual sections as one might expect.

Smeed (1961, p.28) gives a route factor of 1.20 for the central area, confirming the smaller value, although the date of the determination is not given. Smeed had calculated the value by superimposing a square grid of one-sixth of a mile mesh size on the central area and comparing for selected points the shortest distances

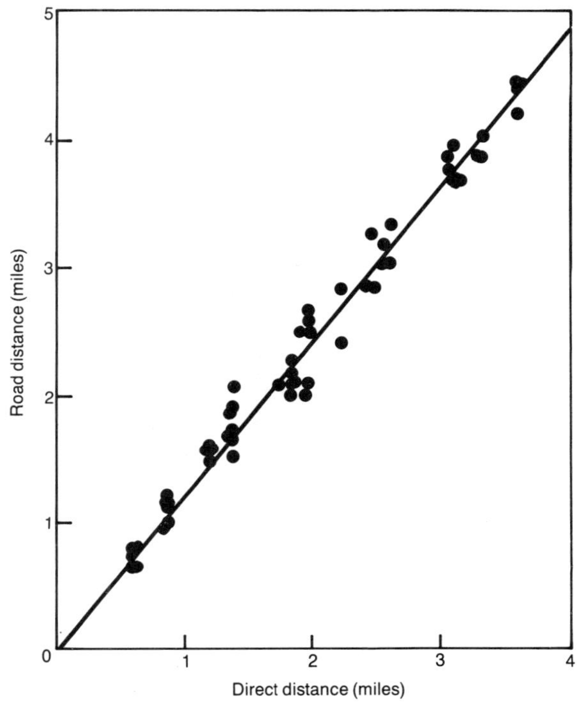

Figure 2.4 Smeed's determination of the route factor for central London

by road with the direct distances. His figure 7 is reproduced below as Figure 2.4.

There was considerable debate in the 60s over the value of traffic management measures which, it was argued, essentially enhanced capacity and increased travel speeds along the road, but did not increase direct speeds. We will come back to this point.

It should be noted that the average direct journey length in this random survey, at 3.4 miles or 5.2 km, is precisely what one would expect from a random trip in a 10-km square. It does not represent what people actually do, as we shall see, being much longer than the actual trips contained wholly in the square.

The variation in travel times on the surveyed routes can be seen in Figure 2.5, which gives the travel times against the direct distances for each section of the tour in 1983.

It can be seen that the variation about the regression line giving average direct speed is very high, with a number of points recording values more than twice as long or half as long as the average time taken. It must be remembered here, however, that different route sections have different route factors which will affect the direct speeds.

Confirmation of the extent of the normal level of variation in central London road journeys can be found in two other surveys, that of the journeys of Smeed from his home in Bray (near Maidenhead) to the Transport Studies Group reported in Smeed and Jeffcoate (1971), and that of the journeys of Fry from her home in Clapham to the Transport Studies Group reported in Mogridge and Fry (1984). Smeed's journeys can be partitioned at the Hogarth Roundabout, giving approximately 7.6 miles of central London journeys (the route taken varied slightly day-by-day), whereas Fry's journeys are 5.8 miles long in the morning and 6.3 miles long in the evening (owing to the one-way systems).

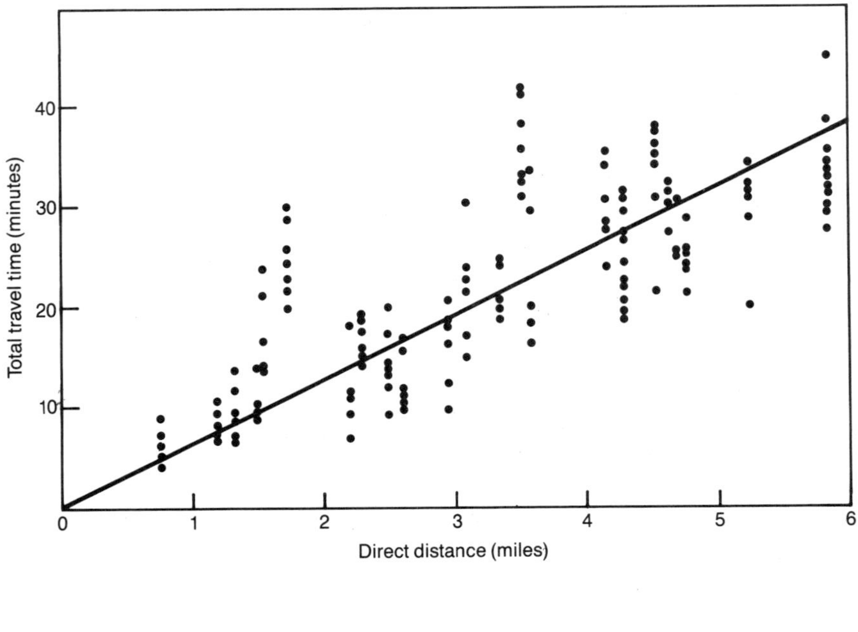

Figure 2.5 Travel time versus direct distance

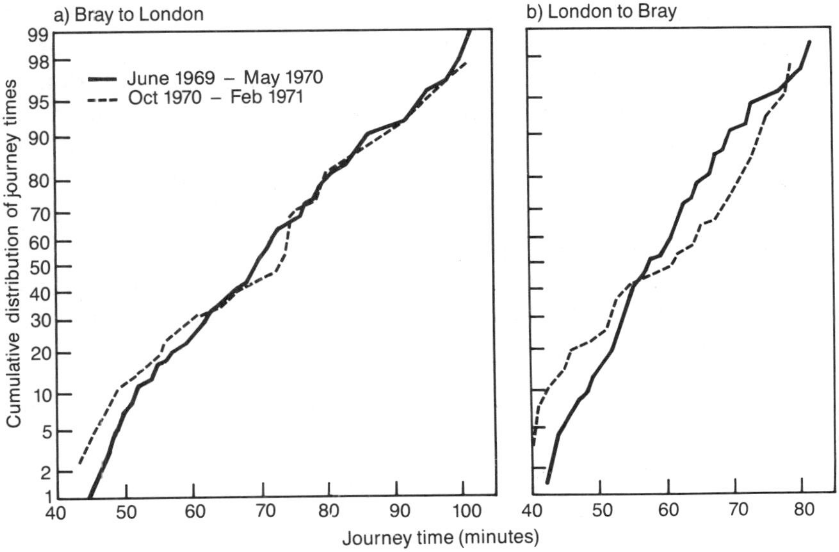

Figure 2.6 Cumulative distribution of travel times on a central London car journey

Smeed's journeys cover two periods, June 1969 to May 1970 and October 1970 to February 1971. In the morning, in the first period the central section took 35.6 minutes and in the second period 32.9 minutes, at speeds of 12.8 and 13.9 miles/h and coefficients of variation of 0.23 and 0.23 respectively. (The coefficient of variation is the standard deviation divided by the mean value and is thus a measure of relative spread.) In the evening, the central section took 34.3 and 32.5 minutes, at speeds of 13.3 and 14.0 miles/h and coefficients of variation of 0.20 and 0.33 respectively.

The number of journeys recorded was 124 and 47 in the morning periods and 73 and 41 in the evening periods respectively, but start times varied from Maidenhead from 6.59 am to 10.50 am and from the Transport Studies Group from 3.23 pm to 10.28 pm, of which about half start between 7.45 am and 8.15 am and between 5.45 pm and 6.45 pm. We are not therefore dealing with midday off-peak journeys as in the random survey, but we are dealing with a sizeable number of journeys.

These speeds are thus approximately the same as those in the random survey noted above, even though the Hogarth Roundabout is further out from the centre than the area covered by the 10-km square. More importantly, we can see that the coefficient of variation is high. Smeed and Jeffcoate presented their results in a Figure where normal distributions of travel speed would be straight lines. Their results for the total journey from Bray (Maidenhead) to the Transport Studies Group are given in Figure 2.6.

Mogridge and Fry give for their journeys a Figure where logarithmic normal distributions of travel speed would be straight lines. Their results are given in Figure 2.7.

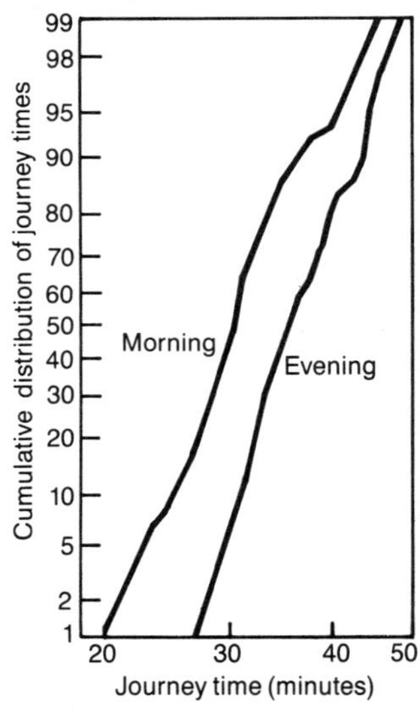

Figure 2.7 Cumulative distribution of travel times on a central London car journey

Although the journeys recorded by Mogridge and Fry are mainly peak-hour journeys, it can be seen that their average road speeds are 11.5 miles/h in the morning peak and 10.3 miles/h in the evening peak. These are slightly lower than the results of Smeed and Jeffcoate because more of the journeys are in the peak hour and all of the journey is in the centre. A slightly lower coefficient of variation of about 0.15 was also found, which is also to be expected if more of the journeys are in the same period of time.

A further set of results, again from Smeed but this time recorded in the '50s and along relatively short one-mile stretches of road in various parts of London with 30 records of speed per stretch (Smeed, 1961), shows that the standard deviation increases as road speed increases but that the coefficient of variation (the standard deviation divided by the mean) decreases as road speed increases. The cumulative distributions of speed are plotted in Figure 2.8 below.

Table 2.2 Means and standard deviations of travel speeds in test runs on a selection of roads (Smeed, 1961)

Mean speed (miles/h)	*Standard deviation (miles/h)*	*Coefficient of variation*
5.7	1.8	0.32
7.1	1.8	0.25
8.2	1.8	0.22
9.2	2.0	0.22
12.8	2.3	0.18
13.3	2.7	0.20
13.4	3.6	0.27
19.7	2.8	0.14
21.9	2.9	0.13

It has thus been known for a long time that there is a high variability in the travel speed on roads in London, and therefore that, if one is to determine an average speed with any accuracy, one has to measure the journey time a sufficient number of times.

In fact Newby in 1949 and Wardrop and Turner in 1951 addressed this problem, as a result of their experience with the first in the series of Road Research Laboratory London Traffic Surveys. These were started in 1947 and are described shortly. Wardrop and Turner estimated that, with the variation observed in the data collected, some 3000km of road needed to be surveyed if an

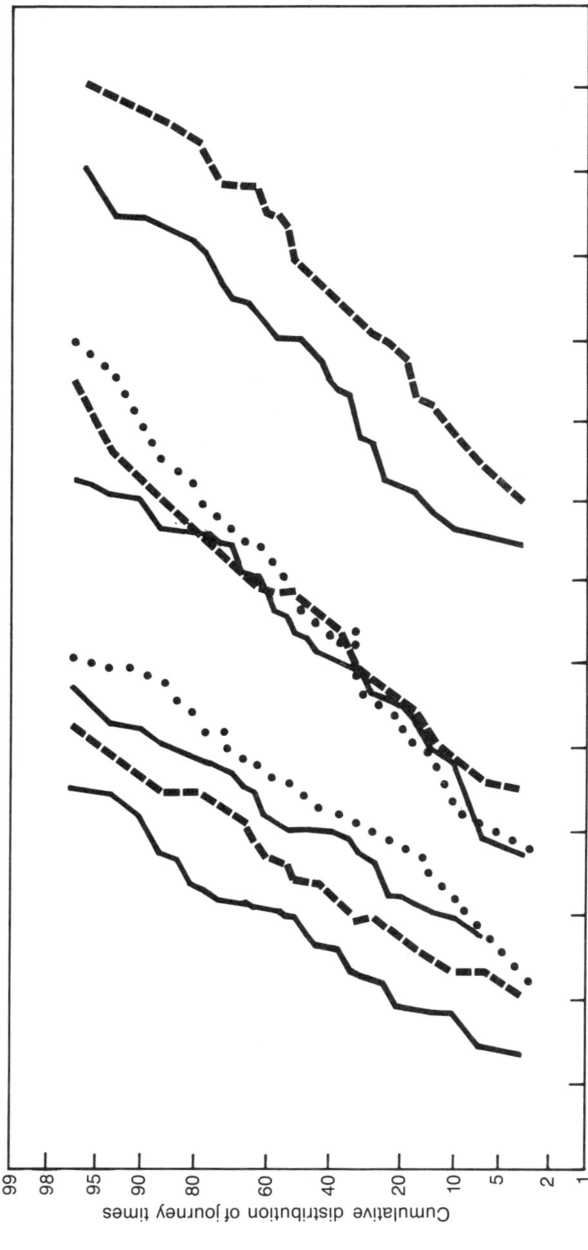

Figure 2.8 *Cumulative distribution of travel speeds on a set of London road stretches of 1 mile in length*

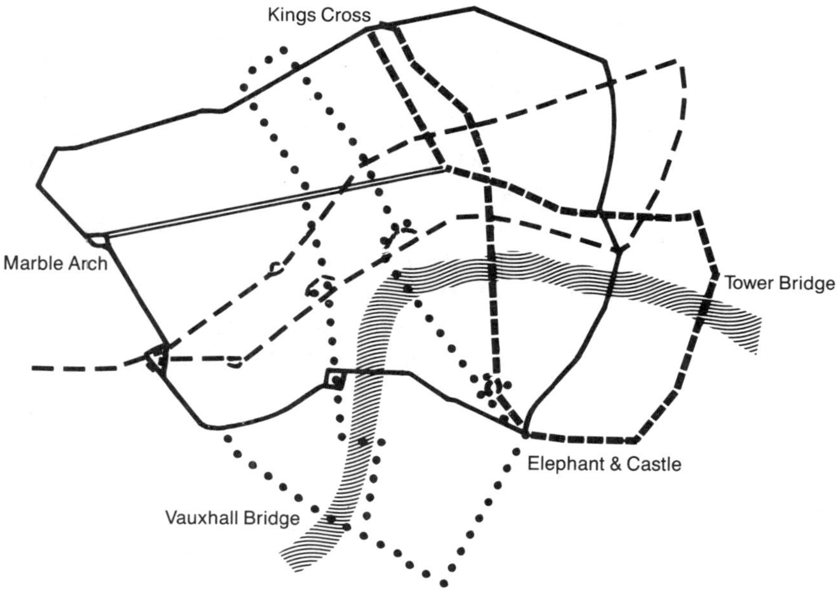

Figure 2.9 The RRL Survey routes in central London

accuracy of 1 per cent in the mean speed was required. This is given in Wardrop's 1952 paper.

No such accuracy had been quoted in any of the official speed determinations until I made some rough estimates for the random speed surveys in my lunch-hour lecture at University College London in 1985 (Mogridge, 1986). Subsequently, in the 1987 report from the Department of Transport which continues the series on London traffic speeds, an estimate of the accuracy has been calculated and is thus available for the first time.

It is particularly noteworthy that in the four Transport Assessment Studies in London set up by the Department of Transport in 1985 (see Chapter Eight) to study the transport problems of various corridors in London, none of the first-phase reports quotes a distribution of travel speed nor a coefficient of variation, although surveys of travel speed were done and maximum and minimum speeds are given. No comparison is therefore possible with previous work to see whether the variation in travel speeds is better or worse—merely the impression is left that the situation is bad.

We start our examination of average travel speeds with the Road Research Laboratory's (RRL) London Traffic Surveys begun in 1947 by Newby and Wardrop. The network is purely central London, and consists of a set of four tours or circular routes of about nine miles each, thus covering about 36 miles in total. Oxford Street was added later in 1949. From 1962 on, extensive changes with one-way systems complicate the route pattern. Each route was surveyed 15 times in the off-peak and three times in the evening peak in each direction from 1952 onwards every two years. This gives an evening peak survey mileage of roughly 200 miles.

The results are as follows:

Table 2.3 Central London traffic speeds, London Traffic Surveys 1947–66 (miles/h)

	off-peak	*peak*
1947[1]	11.1	—
1947[2]	11.5	
1949	11.4	—
1950	10.9	—
1952	11.1	10.8
1954	10.8	10.0
1956	10.4	9.1
1958	10.1	8.2
1960	9.8	8.5
1962	10.4	9.5
1964	10.6	8.7
1966	10.7	9.5

The 1947 values are for a first survey[1] in September and a subsequent second survey[2] in December. The 1966 value was actually done by the newly formed Greater London Council (GLC) and is reported in Turner and Crawford (1966).

It is noteworthy that, following the first survey of the evening peak-hour in 1952, a Pathé News feature, 'The Great Hold-up', in 1953 used the phrase 'Jam yesterday, jam today but please, no jam tomorrow'. I was unaware of this at the time I titled my University College London Lunch-Hour Lecture in 1985 'Jam yesterday, jam today and jam tomorrow?', which is now the subtitle of this book.

Subsequent values of road speeds were obtained by the GLC on a much broader basis covering the whole of the GLC area. The set-up is described in Harrison (1969). The cycle has increased from two years to three but, more importantly, the increase in the coverage has been obtained at the cost of a much reduced frequency of survey. In the centre, the coverage has increased to approximately 100 miles of main roads, including a much better coverage of the West End rather than the emphasis on the City in the east of the centre as before. The number of runs was, however, initially reduced to 1 in each of the morning and evening peaks and two in the midday off-peak (one before lunch and one after).

It is particularly unfortunate that no comparison seems to have been made in 1968 between the values obtained under the old coverage and the new, so that there is a gap in the sequence. In the Department of Transport's (DTp) compilation of historical results (Department of Transport, 1989b), there is likewise no attempt to estimate what the traffic speed values would have been in the latest central area survey in 1986 using the old RRL survey coverage as compared to the value obtained with the GLC coverage. This is no doubt because the road network changes made such a comparison too difficult.

There is, however, another survey which can be used to try to estimate the difference. In an independent measurement of the traffic speeds in central London, the London Traffic Survey (LTS) of 1962 measured a much larger set of roads in the central area, as part of a comprehensive survey of transport in London. This gave an average off-peak value of 12.1 miles/h (LTS, 1964, Table 2.19). Whilst it is not known what the result would have been for the RRL roads surveyed, a value is available for the GLC-surveyed roads which is 12.4 miles/h (Harrison, 1969).

As we shall see shortly, the result for 1968, the first GLC survey, was 12.1 miles/h in the daytime off-peak. We can conclude that

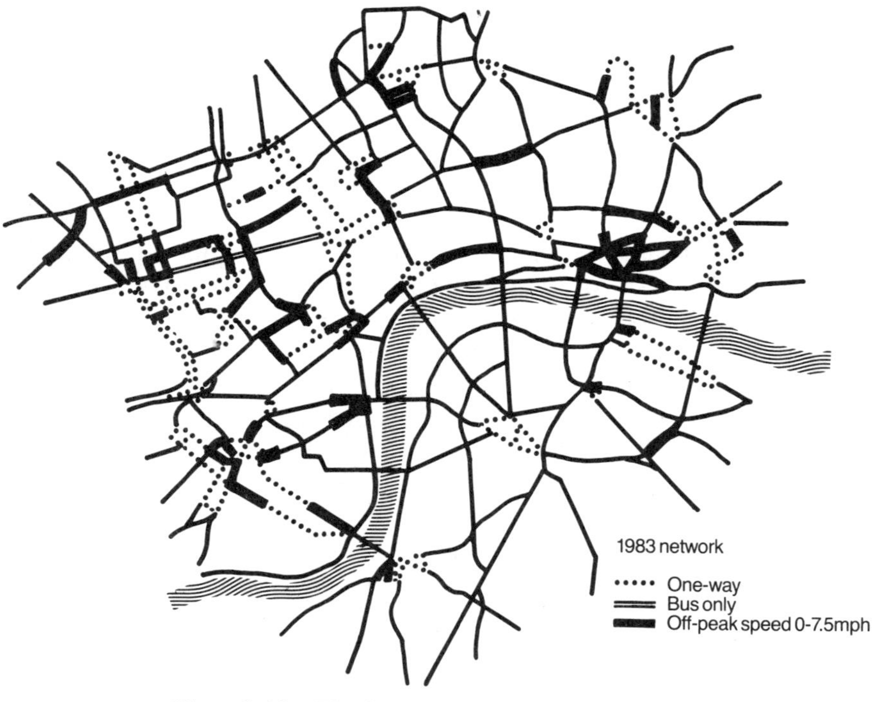

Figure 2.10 The GLC Survey routes in central London

there was no significant change in traffic speed in the six years 1962–68, and that therefore the RRL survey base was giving results about 2 miles/h lower than the GLC base because of its different coverage.

The 1962 LTS also gives for the first time a comprehensive set of values of speed on roads outside the central area. For inner London, an off-peak value of 17.6 miles/h was given; for outer London, an off-peak speed of 25.6 miles/h was given, but this includes a rather larger area than the GLC area. For the total survey area, the average off-peak speed was quoted as 21.5 miles/h.

The random surveys discussed above have no such problem of comparability of roads; we merely take the best route available and measure the change in direct speed.

The results of the GLC surveys, including the continuation by the DTp (on a slightly different cycle basis), are given in Table 2.4.

Table 2.4 London traffic speeds, London traffic surveys 1968–88 (miles/h)

Cycle Number	*1* *1968*	*2* *1971*	*3* *1974*	*4* *1977*	*5* *1980*	*6* *1983*	*7* *1986*
Central London							
Morning peak	12.7	12.9	14.2	12.3	12.1	11.8	11.5
Daytime off-peak	12.1	12.6	12.9	12.6	11.6	11.9	11.0
Evening peak	11.8	12.7	13.2	11.9	12.2	11.5	11.0
Inner London							
Morning peak	14.4	13.8	15.3	13.3	13.5	12.9	11.2
Daytime offpeak	17.4	17.7	17.7	16.4	16.3	15.4	13.4
Evening peak	14.6	13.9	14.8	12.9	13.6	12.5	11.2
Outer London							
Morning peak	19.4	18.7	18.0	17.6	18.0	17.2	
Daytime offpeak	24.6	24.2	23.6	22.6	22.4	22.7	
Evening peak	20.5	20.0	18.7	18.3	18.7	18.2	
All roads							
Morning peak	18.1	17.7	17.9	16.9	17.5	16.9	
Daytime offpeak	21.3	21.6	21.7	20.9	20.6	20.9	
Evening peak	18.6	18.3	18.3	17.2	18.0	17.2	

It can be seen by comparing these GLC results with the LTS results that the off-peak road speeds hardly altered in the period from 1962 to 1971. The 1988 report, by the DTp, is the first to publish an analysis of the standard errors in these results. These are given in Table 2.5 for the sixth (GLC) cycle. The standard error in the seventh (DTp) cycle, given in brackets, is only available for the central area so far.

As can be seen from these errors, it is quite impossible to distinguish from two successive surveys in the central area whether or not there has been a change in traffic speeds; in the inner and outer areas, though, the differences are sometimes large enough to be fairly certain a change has occurred.

The estimated speed on individual stretches of road is, as might be expected, subject to a high sampling error with such a low

Table 2.5 Standard errors in speeds

	Number of road links	*Number of survey days*	*Estimated standard error (miles/h)*
Central area	680(700)	22 (16)	0.45 (0.60)
Inner London	880	60	0.30
Outer London	1410	91	0.25
All areas	2970	179	0.20

sampling rate; the DTp (1989c, p.2) quotes an estimate that the speed will lie within six miles/h of the quoted value in 95 per cent of the cases. Successive surveys of the speed on the same road may thus be 10 miles/h different and still be plausibly within the sampling error. Nevertheless, as we shall see shortly, it is possible to make comparisons over long stretches of route with a reasonable degree of accuracy.

The variation between days on the same road cannot be measured with any accuracy because the sample is not large enough to cover repeated runs on different days on the same road.

The central and inner areas are now covered twice, ie two runs in the morning peak (three hours), two runs in the morning off-peak (10 am–1 pm) and the afternoon off-peak (1 pm–4 pm), making six hours, and two runs in the evening peak (three hours). The measured speeds avoid school holidays and adverse weather conditions; it is also believed that they omit roads whose speed is affected by road-works, breakdowns etc, though this is not stated.

Speeds are flow-weighted but, rather intriguingly, the flows are only collected in the off-peak, and peak flows are estimated from these off-peak flows using peak-hour factors from other surveys. (DTp, 1988a, p.47). All-day (7 am–7 pm) speeds are calculated using ratios of 3:6:3 for the morning peak, off-peak and evening peak raw flow-weighted data respectively.

The results in the 1974–75 period for the third cycle for the central and inner areas are, to quote the DTp (1989c para.27), best seen as exceptions caused by the oil crisis and its aftermath rather than as part of the general trend.

If we perform a linear regression on these sets of results to pick up any time trend (ignoring the third cycle for the central area), we obtain the following set of equations (where t represents the three-year cycle number):

Table 2.6 Regression equations of mean speeds 1968–86 (miles/h)

Central London	Morning peak	13.14 – 0.22 t
	Daytime offpeak	12.75 – 0.19 t
	Evening peak	12.54 – 0.17 t
Inner London	Morning peak	15.37 – 0.47 t
	Daytime offpeak	18.90 – 0.64 t
	Evening peak	15.39 – 0.57 t
Outer London	Morning peak	19.50 – 0.39 t
	Daytime offpeak	24.94 – 0.45 t
	Evening peak	20.65 – 0.45 t
All roads	Morning peak	18.26 – 0.22 t
	Daytime offpeak	21.75 – 0.17 t
	Evening peak	18.83 – 0.26 t

It is obvious from these time trends that traffic speeds in London are essentially very stable, but that there is a slow but significant fall in speeds in all areas and in all daytime periods, being least in the centre and most in the inner area (though most of this is due to the fall in values of speed in the seventh cycle; if we ignore this, the outer area shows the most consistent fall).

A fall of 0.06 miles/h per annum in the central area and 0.15 miles/h per annum in the outer area is hardly fast, but is significant, giving over a twenty-year period falls of 1.2 miles/h and 3 miles/h respectively, which are substantially bigger than the standard errors quoted above for individual years.

The latest inner area results show a much bigger fall than would be expected on the previous trend.

If we perform a linear regression on the RRL results from 1952 to 1966, we obtain the following (where t now represents a 2-year cycle).

Table 2.7 Regression equations of mean speeds 1952–66 (miles/h)

Daytime off-peak	10.71 – 0.05 t
Evening peak	10.04 – 0.17 t

We see that the fall in the evening peak is 0.085 mile/h per year, almost the same as for the later GLC results, but the off-peak fall is negligible at 0.025 miles/h per year. We also see that by these regression estimates the 1966 values are:

Table 2.8 Estimated 1966 mean speeds in RRL and GLC surveys (miles/h)

	RRL	*GLC*
Daytime off-peak	10.31	12.69
Evening peak	8.68	12.48

The estimates of RRL values in 1966 for the evening peak are thus considerably lower than those for the off-peak, whereas the GLC values are very similar. The off-peak, however, has considerably better coverage in the RRL survey than the peak, so is likely to be a more accurate estimate. The differences confirm our earlier estimate that the RRL survey was recording values of speed about two miles/h less than the GLC survey because of its different coverage of roads in the central area.

One must not be misled by these results into thinking that road speeds are falling. What these results show is that *flow-weighted* speeds are gradually falling in each area and in each time period, more in outer London than in central. But it does not necessarily show that speeds on a given journey are changing, if the pattern of travel is changing. That is, there could simply be a greater increase in traffic in those parts of the areas where traffic speeds are slower.

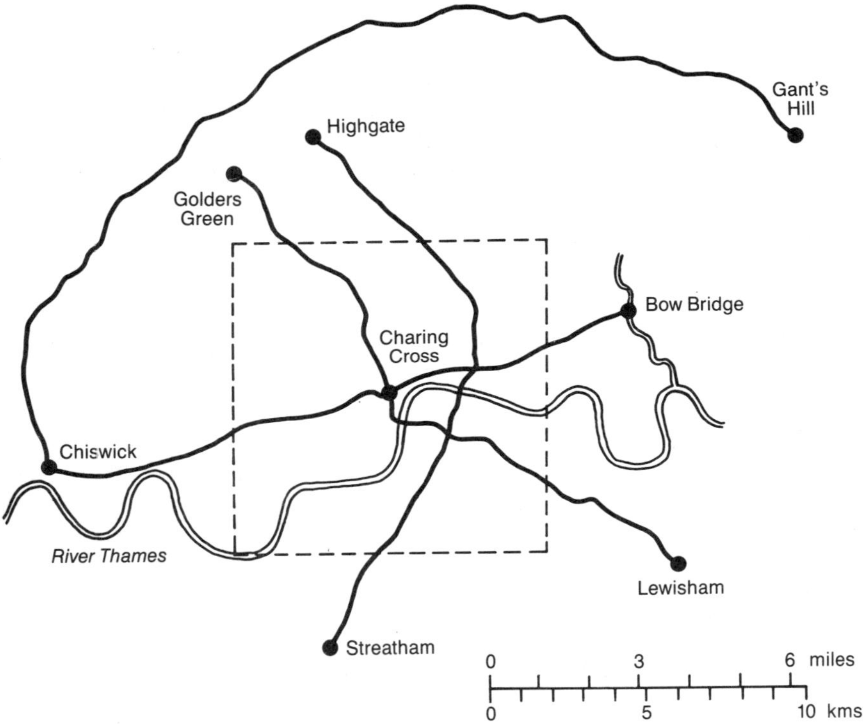

Figure 2.11 Bressey and Lutyens' routes in 1936

To see this we need to examine some other sources of data on traffic speeds. The most illuminating are the surveys recorded in 1936 by Bressey and Lutyens for the Highway Development Survey (Bressey and Lutyens, 1938). Sir Charles Bressey was a former Chief Engineer of the Ministry of Transport and Sir Edwin Lutyens was an architect and planner (of Delhi, for example).

Three routes through the centre of London were chosen, but also one around the nascent North Circular Road. The routes are shown in Figure 2.11.

The routes across the centre are all 12.6 miles long; that around the North Circular was quoted as 22.75 miles. The survey procedure was quoted as follows:

> 'Each of the four routes mentioned was traversed continuously by a 16 hp Austin Light-Six Touring Car during a six-day working week (Sunday excluded). From Monday to Friday inclusive, the car ran from 8 am to 7 pm, and on Saturdays from 8 am to 1 pm. The car was driven by a steady and competent professional chauffeur, who had no inducement to break records or take risks. His pace may be assumed to be that of the punctiliously cautious and considerate driver who presumably constitutes the bulk of the British motoring community. Officers in the car kept stop-watch observations of all measurable delays, and noted the points of occurrence'.

And the results:

> 'The numerous journeys performed on the three cross-London routes indicate an average speed of roughly 12.5 miles per hour from end to end. This can usefully be compared with the North Circular Road, where the corresponding figure is 23.6 miles per hour'.

The Department of Transport (DTp, 1989b para.33), from its latest survey results in 1986–87, was able to compute the values of traffic speed on the same routes. For the three central routes they give 12.4 miles/h; for the North Circular Road they give 23.1 miles/h. These are identical within survey errors to the results 50 years earlier!

This is the more remarkable because the speeds are not the same over all the routes or over all the time periods. In 1936, values of average speed were recorded on the central routes for individual sections from Ludgate Circus to Commercial Road at 5.8 miles/h, and between Euston Road and Trafalgar Square at 7.7 miles/h. The slowest speeds recorded on these sections were 3.6 miles/h and 6.3 miles/h respectively. Speeds on the outer sections would have to be correspondingly higher than average over the route.

In 1986, the North Circular Road values of average speeds for each time period are given as 19 miles/h in the morning peak, 24 miles/h in the midday off-peak, 18 miles/h in the evening peak and 28 miles/h in the evening off-peak. It will be seen that the peak and midday off-peak values are very close to the Outer London average speeds for these time periods; that is, the road has average speeds on it very close to the average speeds for the area through which it passes.

It is worth noting that Newby (1954) attempted to use the London Traffic Surveys to compare those results with the results of Bressey and Lutyens. Newby was, however, limited both by the fact that the routes in the London Traffic Surveys only corresponded over short distances with those of the 1936 survey, and by the fact that the London Traffic Surveys were only repeated 15 times in the off-peak and three times in the evening peak (and none in the morning peak), whereas the 1936 survey was 8 am to 7 pm (and Saturday morning)—although the number of runs was unspecified but 'numerous'.

Table 2.9 Mean speeds in central London 1936 and 1952

Route	*Distance (miles)*	*1936*	*1952*
		(miles/h)	
City Road/Old Street to Elephant & Castle via London Bridge	2.32	8.0	11.0
New Kent Road/Bricklayers Arms to Warren Street via Westminster Bridge	3.27	8.9	9.9
Hyde Park Corner to Piccadilly Circus	0.75	7.0	8.9
Trafalgar Square to Ludgate Circus	1.12	9.8	10.8
Bank to Houndsditch	0.59	8.4	10.1

Newby appeared to think that this increase in speeds between the two surveys was significant, with an average increase over the 8.05 miles of road from 8.4 to 10.1 mph or 20 per cent. There are, however, substantial differences between the two surveys, in the time periods and in particular in the number of runs over a particular route. This is extremely important, given the variation in the journey speed from run to run, and it is unfortunate that the published data do not give either the number of runs or the

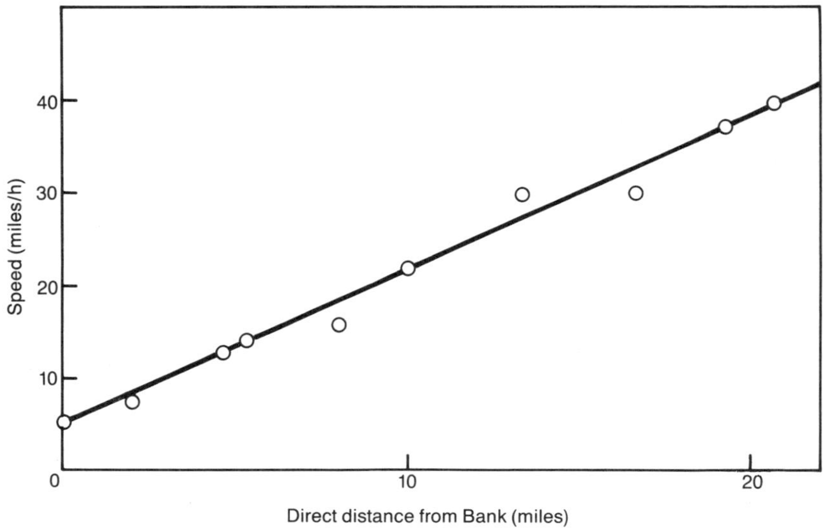

Figure 2.12 Roth's 1959 values of speed (in miles/h)

coefficient of variation in the mean speed to enable a judgement to be made about whether the difference is significant or not.

Another set of travel speeds can be obtained from the runs made by Roth in 1959 during his work on the determination of fuel consumption by different types of vehicles at different speeds. The runs were made on a route from the Road Research Laboratory at Langley directly into and around central London. The number of runs was 17 by car, 17 by van, and 4 by lorry (the vans were Post Office vans!). We plot the values of speed against the distance from Bank in Figure 2.12.

It can be seen from these results, the first to give a route which goes right outside London, that the speed increases from its lowest values in the City to its highest values on the arterial road near to and beyond the border of the Greater London conurbation.

Of particular note is the route from Trafalgar Square to Ludgate Circus, a combination of Strand, Aldwych and Fleet Street, for which Roth gives speeds of 7.8, 11.9 and 10.4 miles/h respectively, which when appropriately weighted by distance give an average speed of about 9.2 miles/h. This is rather less than either of the 1936

(9.8) or 1952 (10.8) values, and suggests that speeds had not, after all, increased as Newby argued.

In Mogridge (1986a), I noted that the measurement of traffic speeds in central London could in fact be taken further back still, to the early days of this century. I noted in particular a comment by the Post Office Committee set up to consider the proposal for the Post Office to build its own railway across and under central London, from Whitechapel to Paddington via Oxford Street. The 1911 report (Postmaster General, 1911) came to the following conclusion:

> 'Whether it will ever be possible to obtain an average reliable speed of over 8 miles an hour during ordinary business hours in central London *even with motor vans* is a matter of extreme doubt' (my emphasis).

The Post Office eventually got Parliamentary permission to build and operate its own railway which, owing to the interruption of the 1914–18 war, was not finally opened until 1927 along the line shown below. The more adventurous proposals to connect all the major rail termini were dropped because of lack of finance.

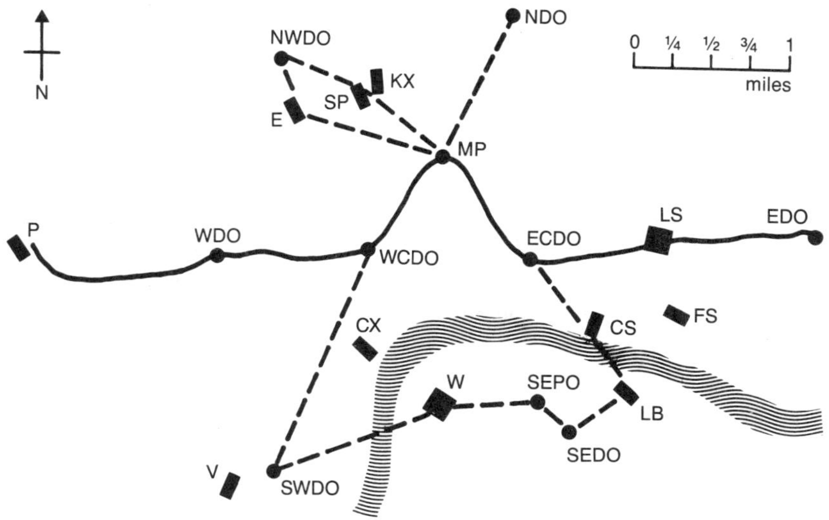

Figure 2.13 The Post Office railway in central London

I examined the data from the Post Office Committee in Mogridge (1988a). Returns taken during two normal days in October 1909 and June 1910 gave the following set of travel speeds between the District Post Offices and/or the railway terminus at Paddington.

Table 2.10 Speeds of Horse-drawn vans between Post Offices in 1909–10 (miles/h)

		return
Paddington, WDO, WCDO, ECDO	6.5	6.5
SWDO, WCDO, ECDO	6.5	
SWDO, ECDO	6.75	
Mount Pleasant, ECDO	5	5.5
Paddington, WDO, WCDO, Mount Pleasant	6.5	6.5
SWDO, WCDO, Mount Pleasant	6.75	
Mount Pleasant, WCDO	6	
ECDO, WCDO	6	
SWDO, WCDO	7	
ECDO, WCDO, WDO	6.25	
Mount Pleasant, WCDO, WDO	6.25	
EDO, ECDO	5.25	

Table 2.11 Speeds of different vehicles along various routes in central London 1908 (miles/h)

(i) Along the Strand from Charing Cross PO to the Law Courts
(ii) Along Oxford Street from Tottenham Court Road to Oxford Circus
(iii) Along Piccadilly from Piccadilly Circus to Half Moon Street

	(i)	(ii)	(iii)
Horse cab	8	7.5	7.25
Horse bus	6.5	6	5
Motor cab	9.33	6.5	7.25
Motor bus	8.75	6.25	5.5

It can be seen that these speeds average about 6.5 miles/h, with those in the City slightly lower.

Another set of observations considered by the Post Office Committee were taken in June 1908 for the Highways Committee of Westminster City Council and are given in table 2.11. It is not stated how many observations were included for each class of vehicle, but it is obvious from the variation that it cannot be stated that motor vehicles are faster than horse-drawn vehicles in the congested conditions of these central London streets.

Recalling Roth's 1959 values, half a century later, of 7.8 miles/h along the Strand and 11.9 miles/h round Aldwych, and appropriately weighting by distance, the comparable speed in 1959 appears to be 9.4 miles/h, very similar to the speed of the motor cab in 1909.

I went on to comment in Mogridge (1988a) that an even more extensive set of traffic speeds for mail vans, both horse-drawn and motor, had been given in evidence to the House of Commons Select Committee on the Post Office (London) Railway Bill on 3 July 1913 by Sir Robert Bruce. These were taken during a fortnight in October 1912 and give speeds for each of the six four-hour periods of the day, starting at midnight, for many of the routes between the District Post Offices and/or Paddington.

Whilst the speed of horse-drawn vans rarely rose above eight miles/h at any time of the day, the speed of motor vans was the same as that of horse-drawn in the most congested areas of the City for most of the day, typically about six miles/h, and only rose above it towards the west (i.e. towards Paddington) and in the slack periods of the day; between midnight and 4 am, the maximum speed of motor vans was achieved on the route between Paddington and the WDO at just over 13 miles/h.

In addition, the speed of motor omnibuses between Kingsway and Liverpool Street Station was quoted at 5.8 miles/h, and along Cheapside itself at 3.25 miles/h. This was from evidence given by Mr Stanley to the Motor Traffic Committee a few days previously.

The Department of Transport has subsequently republished the detailed tables given by Sir Robert Bruce in DTp (1989b) in their own re-examination of historical road speeds. In paragraph 12, they quote Sir Robert's statement:

'During the last twelve months we have very much increased

> the use of motors in London in place of horse conveyance; but we find that although we have provided motors which can proceed at the rate of twenty miles an hour, we cannot get through the crowded thoroughfares of London at a greater speed, during the business hours of the day, than five, six, seven or eight miles an hour'.

The Department of Transport has attempted to estimate the 1986 speeds by car on the routes traversed by motor-driven Post Office mail vans in 1912. It says, in paragraph 30, that the original Post Office routes are not known, although it is presumed they were the simplest direct routes available, and the best comparison that can be made with current traffic speeds is on sensibly selected equivalent routes.

In paragraph 31, it says that the comparisons are believed to be slightly flattering to the older surveys. The modern results, made up piecemeal fashion, are likely to include rather more time at junctions spent turning onto and off the routes than the original runs, where the usual choice of exit at junctions would be straight ahead.

The results for the comparison over the whole set of routes in 1986 give 10.4 miles/h in the morning peak, 9.3 miles/h in the daytime off-peak and 9.8 miles/h in the evening peak periods. The original Post Office studies were conducted in four-hourly periods beginning at midnight, so that the closest comparable periods of the day are 8 am to 12 noon, 12 noon to 4 pm and 4 pm to 8 pm; these give 6.7 miles/h, 7.3 miles/h and 7.3 miles/h respectively. Over the whole working day, the 1912 average was therefore about 7 miles/h, the 1986 average about 10 miles/h.

It concludes that modern traffic speeds are somewhat higher than in the early part of this century over the area covered by these routes; that is, from Paddington to Whitechapel.

If, however, we were to take those areas of central London, particularly in the City, where the speeds of horse-drawn vans in 1912 were equal to those of motor-drawn ones (i.e. where congestion set the limit to the speed), then I surmise that the conclusion would be that modern speeds would still be identical to those of 1912.

The map of road speeds in 1986, in the midday off-peak, shows

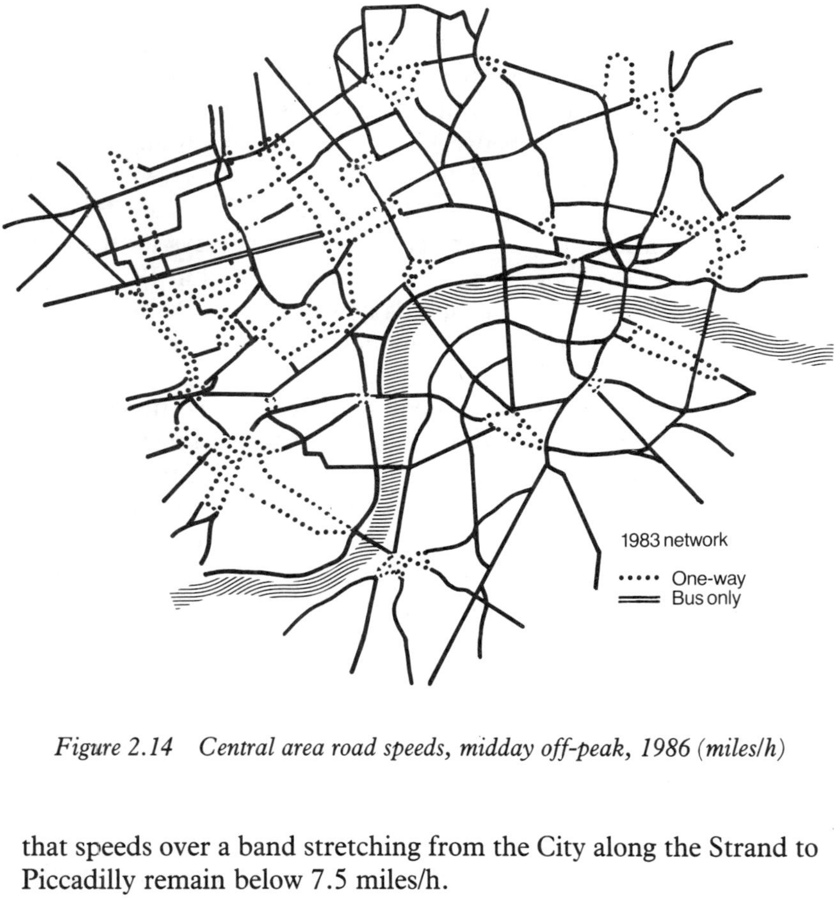

Figure 2.14 Central area road speeds, midday off-peak, 1986 (miles/h)

that speeds over a band stretching from the City along the Strand to Piccadilly remain below 7.5 miles/h.

This, as can be seen by comparison with the previous figure, is roughly the central portion of the route of the Post Office Railway.

When traffic speeds were measured for the Royal Commission for London Traffic in 1904, in Vol. 8 Table 5 (the report of the Advisory Board of Engineers) we find that the speed of horse omnibuses on a route between Clarges Street, Piccadilly, and Bank averaged 5.48 miles/h over 105 journeys between 8 am and 9 pm, and 5.27 miles/h in the reverse direction with the same conditions.

The City Solicitor, Sir Homewood Crawford, in his evidence (Vol. 3, Appendix 79, Table B) had given the times for each of 10 runs on each of three days for five omnibus routes across the City, and had found speeds which varied over the range of 3.5 to six miles/h in the 'Crush Hour' and from 5.75 to eight miles/h in the 'Slack Hours' for horse omnibuses and 6.5 to 8.25 miles/h and 8.25 to 11.5 miles/h respectively for motor omnibuses. The speeds for horse-drawn cabs over four routes across the City, with the same other conditions, varied between 3.5 and six miles/h and six and 8.5 miles/h respectively, and for motor cabs were given as eight miles/h in the 'Crush Hour' and 12 miles/h in the 'Slack Hours'. It is my contention that, if repeated now, these surveys for the City of London and the congested parts of the West End in 1904 would give the same results for road speeds now as then, notwithstanding the change in motive power from horse-drawn to motor-driven vehicles.

It most certainly is not my contention, however, that these speeds are immutable, because there is evidence that road speeds were in fact lower in the last century. The first timed runs of traffic speeds that I know of in the City of London were carried out by William Haywood, then the Surveyor of the City Corporation, in 1863.

The mean of from 35 to 69 observations on different routes using horse-drawn omnibuses and four-wheel carriages gave 3.28 miles/h to 4.55 miles/h and was reported to the Select Committee considering the impact of new railway construction (Vol. viii QQ 1226–8) (and is reported in Kellett (1969, 1979, p.50)). This was part of an early attempt to measure the social costs of congestion to which we will return later.

Thus in the period between 1863 and 1904, I believe that road speeds improved substantially in the congested areas of the central area, but they have not improved since then. To see why this might be so we need to examine journey speeds, and this is the intention of the next two chapters.

3 Journey Speeds: Potential Journeys

I mentioned in Chapter 2 that Holroyd and Scraggs' survey in 1962 of road speeds round a tour of 25 random points included journeys by bus, and that I added a pilot study of journeys by rail in 1983. Unlike car journeys, which are kerb-to-kerb and thus do not include access, bus and rail journeys starting from a random set of points defined as road junctions do in practice represent potential journeys door-to-door including access.

This chapter is mainly concerned with presenting the analysis of these potential journeys by bus and rail, but we add at the end what little was known about access times by car in the central area. The next chapter will examine the actual journeys of travellers by car, bus and rail, and give a better definition of access times by car.

One journey between the 25 random points is so short, however, that it is entirely a walk journey, so the number of bus and rail journeys is 24 not 25.

Bus journeys involve walking at each end of the journey and maybe between interchange stops, as well as time spent waiting off the bus for the service and at the interchange if any, plus time spent waiting on the bus at intermediate stops for other passengers to get on or off the bus. One therefore expects bus journeys to be much slower than car kerb-to-kerb journeys.

The following table shows the data for the random bus journeys in the five surveys analysed so far (Ferguson (1985), Heydecker and Ferguson (1990)). There was no bus survey in 1969. Unlike car journeys, where some route variation has occurred because of traffic management, bus journeys have been affected by changes of routes by buses as well.

There is a missing walk journey time in 1988, so that the means are not strictly comparable.

Table 3.1 Central London random journey data for bus

	1962	1972	1977	1983	1988
				minutes	
Mean total time	42.5	47.3	54.0	51.6	55.7
Mean in-bus time	27.2	29.4	32.2	30.4	34.4
Mean walk time	10.6	12.6	11.7	11.8	10.9
Mean wait time	4.7	5.3	10.1	9.4	10.2
				miles/h	
Mean direct speed	4.9	4.4	3.8	4.0	3.6

If we compare the in-bus travel times with those of the car, which were 19.5, 17.6, 20.8, 21.6 and 22.9 minutes respectively, we can see that average bus running speeds are approximately 50 percent slower than car, owing mostly to the fact that they have to stop every

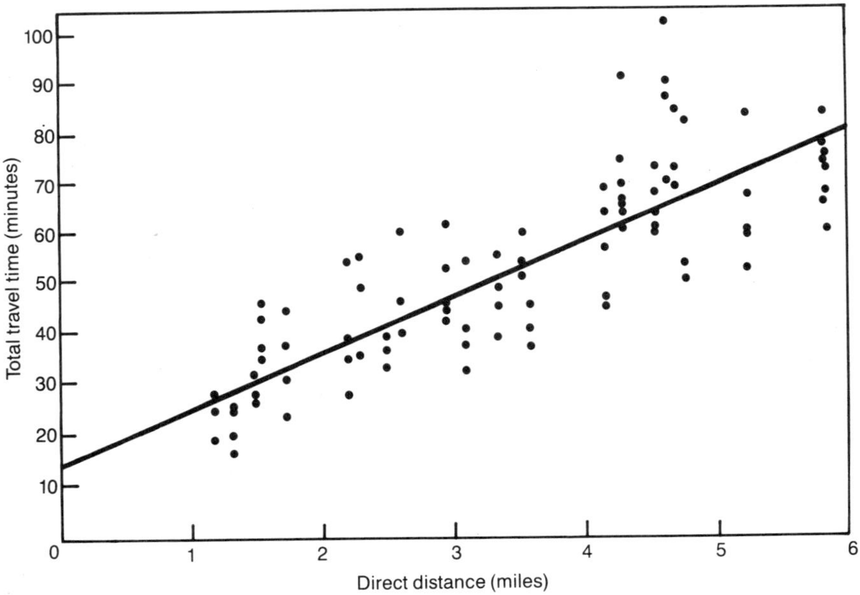

Figure 3.1 Travel time versus direct distance for buses (1983)

so often for passengers. In addition, the walking and waiting adds a further time penalty, so that buses overall, in central London, only give direct journey speeds of some four miles/h, well under half the kerb-to-kerb direct journey speeds of cars.

One particular point to notice from the five surveys is that bus waiting times seem to have increased substantially in the period 1972–77. This was a period when the bus service suffered from a severe shortage of staff, and thus cancellations of services.

The variability of bus journey times with the direct distance of the bus journey can be shown in a similar way to that shown for car journeys in Figure 2.5.

A regression line for these results does not go through the origin as for car journeys, but that for in-bus travel time does. The sets of results are as follows:

Table 3.2 Regression lines for bus journeys

(a) in-bus travel time b (minutes) versus direct distance d (miles)	
1962	b = 8.1d
1972	b = 8.8d
1977	b = 9.5d
1983	b = 9.0d
1988	b = 10.6d
(b) total travel time t (minutes) versus direct distance d (miles)	
1962	t = 6.6 + 10.5d
1972	t = 8.4 + 11.3d
1977	t = 8.6 + 13.3d
1983	t = 13.7 + 11.0d
1988	t = 11.9 + 13.4d

It can be seen that buses have become a little slower over the period, with in-bus direct speeds falling from 7.4 miles/h in 1962 to 5.7 miles/h in 1988.

The regression lines for total travel time show, however, that the total walking and waiting are related to the distance travelled, as one might expect, so that the estimated time taken at zero distance is far less than the mean walk and wait time.

It is not possible to discern a definite trend here, because of the variability of travel times and the inadequacy of the sample size, though I would expect both the intercept time (the time at zero distance) and the slope of the regression line (an inverse of the speed) to rise as the direct speed is falling with time.

The 1983 results have been split into their component parts and regressions of time t (minutes) against the direct distance d (miles) obtained for each component.

Table 3.3 Regression lines for bus journey components (1983)

		R^2
Running (no intercept)	t = 8.96d	—
Running (with intercept)	t = 1.55 + 9.35d	0.80
Walk	t = 12.59 – 0.23d	0.0
Wait	t = 3.49 + 1.72d	0.11
Total	t = 13.66 + 10.99d	0.68

The values exclude a short journey entirely by walk (except that it is included in walk). This means that the values are not entirely consistent. It can be seen that walk is practically independent of direct distance; that is, there is no correlation between the length of the walk stages and the direct distance travelled. This would not necessarily be true of journeys that were much longer, as we shall see.

On the other hand, the length of waiting time increases with distance, mainly due to an increasing number of bus stages in the journey. The correlation is again poor, however.

If we examine separately the one and two-stage journeys in 1983 and 1988 in Table 3.4, we can see this quite clearly. Two long journeys are excluded from this table in 1983 owing to particular problems which affected the results; the results are thus not strictly comparable with Table 3.1. The 1988 results had a number of problems, including a demonstration which delayed one bus,

Table 3.4 Means of bus journey time components (minutes)

	1983		*1988*	
	1 stage	*2 stage*	*1 stage*	*2 stage*
walk to first stop	5.5	5.1	5.1	5.1
walk to second stop		2.6		2.1
walk to end point	5.8	3.7	5.3	3.6
total walk	11.3	11.4	10.0	10.7
percentage of total travel	28.0	21.5	23.7	16.9
wait at first stop	5.9	4.9	6.9	7.3
wait at second stop		5.0		5.6
total wait	5.9	9.9	6.9	12.9
percentage of total travel	14.6	18.8	16.2	20.3
first stage in-bus	23.2	15.8	25.4	16.9
second stage in-bus		15.9		23.0
total in-bus time	23.2	31.7	25.4	39.9
percentage of total travel	61.2	59.7	60.1	62.8
total travel time	40.4	53.0	42.3	63.5

diversions due to collapse of a building during demolition, one bus broke down, several buses were turned short of their destination, extra waiting time due to crew change-over and even an argument between a driver and a conductor! This either added to in-bus time or waiting time.

It can be seen that walking to and from an origin is about five minutes, whereas walking between two bus stops is two or three minutes. This makes total walking a little longer if there are more stages. On the other hand, waiting for each bus is about five to seven minutes irrespective of the number of stages. If only one stage is required, then in-bus time is about 23–25 minutes, whereas if two stages are required, each took about 16 minutes in 1983, a little shorter per stage, but in 1988 the second stage was longer, probably because of the problems noted above.

If we obtain separate regressions of in-bus time b and total time t (minutes) against the direct distance d (miles) for one and two-stage journeys, we find the following:

Table 3.5 Regressions lines for bus journey stages (1983 and 1988)

one-stage journeys	(1983)	b = 9.7d
		t = 13.3 + 11.0d
	(1988)	b = 11.5d
		t = 15.9 + 12.2d
two-stage journeys	(1983)	b = 8.7d
		t = 13.8 + 10.7d
	(1988)	b = 9.8d
		t = 10.2 + 12.8d

Interestingly, these results are very litle different from those for the journeys taken together, although this is slight evidence that the two-stage journeys have a higher in-bus direct speed (6.9 as opposed to 6.7 miles/h in 1983; 6.1 as opposed to 5.7 miles/h in 1988).

Estimates of the time spent waiting for buses started with a study of the pattern of service on a particular route in 1974 (GLC, 1976). This was on route 14, a cross-central route from south-west to north-east. A sample of the tracks of individual buses are shown in Figure 3.2.

This clearly shows the bunching that occurs in conditions of congested traffic. The bus in front becomes slower because of the heavy traffic, picks up more passengers than it otherwise would and becomes even slower, while the bus behind becomes faster as it has fewer passengers to pick up and put down. Eventually the two buses come together and then remain stuck so, because whichever bus is in front has more passengers whereas whichever bus is behind has fewer.

The most buses I have ever seen stuck together in such a way is seven on a nominally three-minute service, but it is fairly common for two or three to be bunched together.

This survey of route 14 buses has since been repeated for other buses such as the 31 in 1981 (London Passenger Transport Research

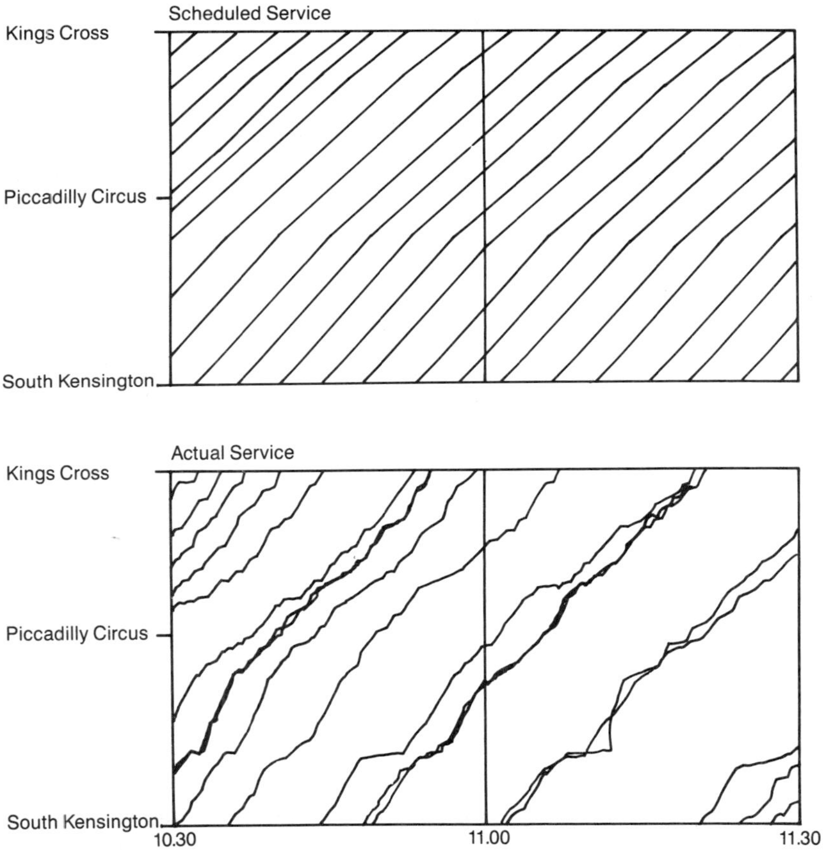

Figure 3.2 Tracks of individual buses along Route 14, 1974

Group, 1983), and roughly the same pattern was found. Interestingly, for this orbital route around the centre, the average waiting time was about 10 minutes.

The tracks of buses in a survey like this can be used to estimate the average waiting time on the basis that passengers arrive at random at each bus stop. If buses were perfectly regular (i.e. arrived at a fixed headway) then passengers would wait on average exactly half the headway.

Table 3.6 Average waiting time versus scheduled headway

Sections of route 14	*Average scheduled headway (mins)*	*Average waiting time (mins)*	
		actual	*perceived*
1	4	3.3	9.9
2	3	4.0	9.5
3	3.5	4.0	8.2
4	3.75	3.7	8.7
5	4.5	5.7	10.2
6	4.5	5.6	10.1
7	4.5	5.6	10.4
8	9	8.4	14.0
9	3.75	5.5	10.4
10	2.75	3.9	10.3
11	3.33	3.7	10.5
12	3.5	5.0	11.8
13	4	6.7	14.4

Bus reliability study Table 3.3

In practice, as the route 14 study showed, passenger average waiting times can often exceed the headway, sometimes substantially. More interestingly, when passengers were asked for their perceived waiting times, they tended to over-estimate the time substantially. The results for actual and perceived waiting times for sections of route 14 are given in Table 3.6.

The sections of route 14 refer to parts of the route, and parts of the day, where different patterns of service are provided. The sample size of respondents in each group was about 150.

It can be seen that the average waiting time was, more often than not, greater than the headway, especially for the higher frequency of service. These average waiting times on a central trunk route in 1974 do, however, lend support to the values obtained in the random surveys in Table 3.4. of five minutes per stage, on stages which were not necessarily trunk routes.

A mean wait time of 5 minutes, however, as in Table 3.1 for 1962 and 1972, which includes two-stage journeys, implies a considerably greater regularity than was occurring in the later surveys of 1977, 1983 and 1988. It is not clear whether this was merely a statistical quirk, due to the small number of journeys in the earlier surveys, or was a real effect. I shall present some more data on this for the peak-period later (see pps 129–130).

It may also be noted that the most efficient operation of a public transport service occurs when the time for the walk to and from the service equals the wait for the service; that is, that there is a strong relationship between the network frequency and the network spacing (TRRL, 1980, Appendix 8, pp. 293–94). It can be seen in Table 3.1 that this condition did not hold in 1962 and 1972, with a walk time about twice the wait time of about five minutes. It did hold in the latter three surveys in 1977, 1983 and 1988 when wait times increased to 10 minutes. Strictly speaking, however, the efficiency criterion is for a single route, rather than a number of stages.

The data in the route 14 survey can also be used to show the variation in bus journey times between the ends of the route. As might be expected, the coefficient of variation is somewhat higher than for cars, and is about 0.20. This is due to the additional variation caused by picking passengers up and putting them down. It is not therefore surprising that, on a high frequency route of any substantial length, with this level of coefficient of variation, buses almost invariably form bunches by the time they reach the end of the route.

Estimates of the time spent waiting for buses have been made on a regular basis by the London Transport Executive (LTE) since 1979. Initially such estimates were system-wide rather than separated for the central area. The 1982 Annual Report (LTE, 1982) quotes a value of 7.5 minutes for average waiting time for the system, based on actual headways and assuming passengers to arrive at random. Values since 1979 have fluctuated slightly from year to year, but remain at about this level.

Results are now available (London Regional Transport (LRT), 1989) which separate off the central area, and which distinguish between high frequency, non-timetabled services (in general every 12 minutes or less) and low frequency, timetabled services.

Interestingly, there is no difference in average waiting time for high frequency services between the central area and the system average. Both are, in 1988, given as 7.2 minutes. The difference is in the scheduled wait; that is, if all buses arrived on schedule. For the central area, the scheduled wait would be 4.0 minutes or half the average headway, whereas for the system it would be 4.5 minutes. The excess represents the effect of variation due to traffic congestion, staff shortages, mechanical failure etc.

The chance of waiting a given time for high frequency buses in the system is quoted as 74 percent for less than 10 minutes, a further 21 percent for less than 20 minutes, a further 4.3 percent for less than 30 minutes, and 1.3 percent for over 30 minutes. For low frequency, timetabled buses, the results are given in a slightly different way. The chance of a bus departing 'on time' (i.e. within two minutes before to five minutes after scheduled to depart) was 64 percent in 1988. Seven percent of buses departed early, and 18 percent were between five and 15 minutes late, with 11 percent more than 15 minutes late or not arriving at all.

These LRT results substantiate the waiting time results already given for the random journey surveys.

It is extremely difficult to find historical data to test whether bus running speeds have changed over time. I reported what I could find in Mogridge (1988a) and the Department of Transport (DTp, 1989b) has subsequently republished the appropriate tables. As we saw in Chapter Two, in 1904 along some routes in the City of London in the 'Slack Hours' motor buses were achieving speeds of between 8.25 and 11 miles/h. Table 3.2, with direct bus running speeds of around seven miles/h, implies a running speed along the route of about 11 miles/h in the off-peak (a slightly higher route factor than for cars as the network of bus routes is more widely spaced) but over a far wider area than the City of London.

On the other hand, the GLC Development Plan Report of Studies (1968) quotes a central bus running speed over the whole day of eight miles/h. This is more akin to the 1904 'Crush Hour' speeds for motor buses of between 6.5 and 8.25 miles/h. Since the frequency of buses in the peak hours is rather higher than in the off-peak, a whole day value can be expected to be nearer the peak hour value than the off-peak value.

Whilst it is not possible to draw hard conclusions, therefore, it does seem as if it would be reasonable to surmise that bus running speeds have not changed much if at all in the centre over this century. This is what we would expect if there had been no change in road running speeds, given that the difference for buses is that they have to stop for passengers every so often and are thus slower.

It must, however, be acknowledged that anecdotal evidence suggests that, since the introduction of driver-only buses in the central area after 1983, the amount of time taken to pick up passengers has increased substantially. This will have caused running speeds to decrease in the last few years. LRT have increased their scheduled running time for driver-only buses by 20 percent compared to driver and conductor-operated buses.

When I became involved in the random survey in 1983, I asked why a comparable rail survey had not been included. Although by then it was too late to include the rail survey in the students' programme (which was, as always, scheduled for March), I carried out the first rail survey myself in August 1983—on the supposition that the rail service does not change with the time of year. The survey was subsequently included in the 1988 programme.

As with the bus survey, only 24 journeys were undertaken, the other being entirely by walk. Two consecutive journeys included a short bus stage, which is treated as if rail. Whereas there are very

Table 3.7 Central London random journey data for rail (1983 and 1988)

	1983		*1988*
		minutes	
total travel time	42.1		42.0
mean in-train time	15.4		14.9
mean walk time	19.2		20.4
outside station		12.7	
inside station		6.5	
mean wait time	7.4		6.6
		miles/h	
mean direct speed	4.9		4.9

few bus journeys requiring three stages, there are few rail journeys only requiring one stage, given that the network is sparser.

The components of the average rail journey (including the walk journey) are given in Table 3.7.

Thus for rail journeys, 35–37 percent of total travel time is in the train, 45–49 percent is walking (of which 15 percent is inside the stations) and 16–18 percent is waiting. This walking outside the stations (12.7 minutes) is slightly greater than that for bus journeys given in Table 3.4. (approximately ten minutes) but there is much more walking involved inside the stations than between buses.

The relation between total travel time and direct distance for rail journeys is given in Figure 3.3.

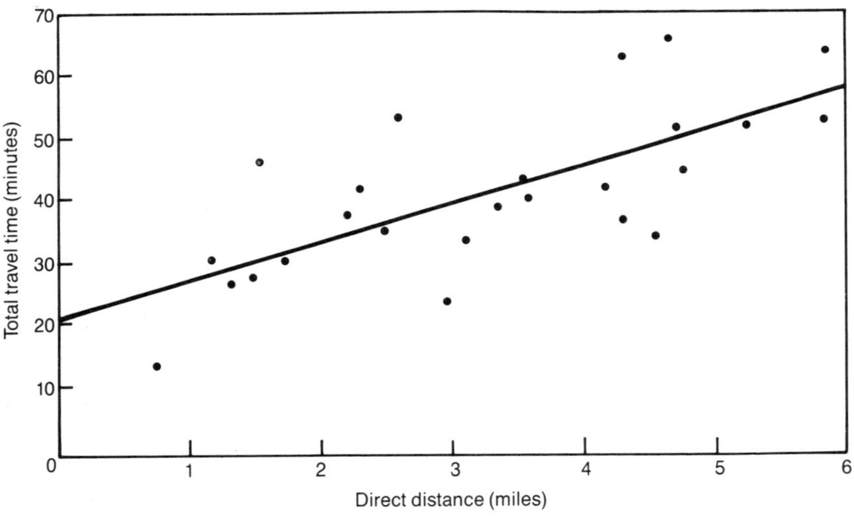

Figure 3.3 Travel time versus direct distance for rail journeys (1983)

The regression analysis of the 24 rail journeys is given in Table 3.8 where r is in-train time (in minutes), t is total travel time (in minutes) and d is direct distance (in miles). The average direct running speed for rail is thus 13.6 miles/h in 1983 (14.0 in 1988), corresponding to a running speed along the route of 18.5 miles/h (19.0) with a route factor of 1.36 (slightly greater than that for

Table 3.8 Regression lines for rail 1983 and 1988

in-train time	(1983)	r = 4.4d
	(1988)	r = 4.3d
total travel time	(1983)	t = 23.6 + 5.4d
	(1988)	t = 22.4 + 5.8d

Table 3.9 Means of rail journey time components (minutes) (1983)

	1 stage		*2 stage*		*3 stage*	
number of journeys	*3*		*11*		*10*	
	1983	*1988*	*1983*	*1988*	*1983*	*1988*
walk to 1st platform	6.0	4.7	8.5	9.0	8.8	9.5
walk to 2nd platform			3.2	2.8	1.8	2.8
walk to 3rd platform					1.2	2.3
walk to endpoint	7.6	10.5	7.2	9.0	9.4	9.6
total walk	13.6	15.2	18.9	20.6	21.2	24.2
percentage of travel time	47.9	49.9	49.7	50.1	41.7	47.3
wait at 1st platform	0.9	5.1	2.9	2.9	4.0	2.8
wait at 2nd platform			3.6	2.9	3.0	2.0
wait at 3rd platform					3.5	4.5
total wait	0.9	5.1	6.6	5.8	10.5	9.2
percentage of travel time	3.2	17.9	17.4	14.0	20.7	17.9
in-train stage 1	13.9	10.1	7.4	7.6	4.9	6.3
in-train stage 2			5.1	7.2	7.9	6.4
in-train stage 3					6.3	5.0
total in-train time	13.9	10.1	12.5	14.8	19.1	17.8
percentage of travel time	48.9	33.2	32.9	35.9	37.6	34.7
total travel time	28.4	30.4	38.0	41.1	50.8	51.2

car in 1983 though comparable to those measured in 1972, 1977 and 1988 in Table 2.1).

The total travel time equation shows, however, just as for bus, that walking and waiting increase with distance, since the average direct running speed in the equation is 11.1 miles/h in 1983 and 10.3 miles/h in 1988. The intercept time of 23.6 minutes is much higher than that for bus of 13.7 minutes in 1983 and 22.4 is higher than 11.9 minutes in 1988 (Table 3.2). Given the scatter of values in the diagrams of journey time versus direct distance, not too much reliance should be put on these values.

The components of rail journey times for journeys by number of stages are given in Table 3.9. It can be seen that the number of journeys of only one stage is really too low to allow any statistical validity to the results, especially to the waiting time which is very low in 1983 and high in 1988.

It can be seen that the walking at each end of the journey is slightly greater than that for bus journeys as one would expect from the sparser rail network, but there is much more walking involved inside the stations. Waiting for service is much less, however, due no doubt to the regularity of service.

The in-train time is considerably less than that for bus, due to the higher speed. In total, rail journeys (for these random journeys) are much faster than those by bus. One can see why I felt I was jumping in and out of trains all day during the survey in 1983; the in-train time per stage hardly gives one time to sit down before one is up again and on the move.

Regression of the components of rail times t (in minutes) against direct distance d (in miles) gives the following, Table 3.10.

Table 3.10 Regression lines of rail time components (1983)

		R^2
running (no intercept)	t = 4.38d	
running (with intercept)	t = 2.04 + 3.86d	0.89
walking	t = 15.36 + 1.08d	0.0
waiting	t = 3.41 + 1.13d	0.09
total	t = 20.81 + 6.07d	0.50

There are small increases in walking and waiting with direct distance (probably in large part owing to extra walking and waiting between stages; there are 2.1 stages in the average journey). These journeys include the one journey made entirely on foot, and two with one bus stage. These bus stages are counted in the running time in analysis, and the bus stops treated as rail stations.

These results may be compared with the desk-top study of Daspal and Gopalakrishnan (1969). These authors used a sample of 50 trips from LT's records (the 1965 Traffic Test data) of tickets sold between the 55 stations of the LT defined central area (that is, the sample is flow-weighted by tickets sold). The journey times are station entrance to station exit. The components of journey time (entry and exit time, waiting time, interchange walking time and running time) are all as given by LT records.

The regression lines between these times t (in minutes) and the direct distance d (miles) are as shown in Table 3.11.

Table 3.11 Regressions of components of time against direct distance for rail journeys (1965)

		R^2
running (no intercept)	t = 4.86d	
entry and exit walking	t = 6.11	
interchange walking	t = 1.53	
waiting	t = 0.92 + 0.49d	0.15
total	t = 8.83 + 5.15d	0.41

There is some inconsistency in these results, as total time is not the sum of the component parts (non-significant variables having been dropped and the relations refitted). The route factor is 1.28.

One has to be careful in comparing these results of a sample of actual journeys in 1965 with those of the random survey of potential journeys in 1983. The average journey in this sample of central London is only 1.6 miles, a journey taking 17 minutes, as compared with the random journeys in the 10-km square which take 42 minutes.

Average direct running speed is 12.3 miles/h, a little slower than the result for the random journeys of 13.6 miles/h, as one might expect if central area stations were a little closer together than in the rather larger area of the 10-km square.

Since the average number of stages in the central area study is not given, one cannot directly compare the walk inside stations of the random survey (6.2 minutes) with that here (6.11 + 1.53 = 7.64 minutes), but one would expect stations in the central area to have slightly longer access and egress times than those of the larger area.

The waiting time regression is quite different, giving an average of 1.7 minutes in the central area study compared to the random survey values of 7.4 minutes in 1983 and 6.6 minutes in 1988 for 2.1 stages, an average wait per stage of 3.5 and 3.1 minutes respectively. It will be noted that such a low waiting time is derived from scheduled trains, not from an actual survey. Rice (1974, p.45) suggests an increase of 20 percent over half the headway, based on a study of morning peak inbound flow on one LT line in 1972 and estimating the variance of headways.

The estimates made by LRT (LRT, 1989) show that average waiting time for rail services was 3.5 minutes in 1988, with 2.7 minutes being the wait if schedules were kept, and 0.9 minutes being the excess waiting time because they are not. These results are system-wide, and include the Docklands Light Railway. 77 percent of passengers had to wait less than five minutes, a further 17 percent less than 10, 4.1 percent less than 15 and 1.7 percent over 15 minutes. The central area is not given separately, only lines. Line data show that the services with the highest frequencies have the largest excess waiting time proportion. The Victoria, with a scheduled wait of 1.7 minutes, has an excess wait of 1.0 minutes. The Bakerloo, with a scheduled wait of 1.8 minutes, has an excess wait of 0.8 minutes.

This system-wide, all-day estimate of 3.5 minutes must be higher than the values one would expect in the central area. It will, however, be remembered that the random survey is an off-peak survey, when train frequencies are lower, and also includes some British Rail (BR) services, which tend to operate at lower frequencies in any case. The random survey values of waiting time therefore seem to be supported by the LRT data.

Some idea of the variation in rail journey times can be obtained

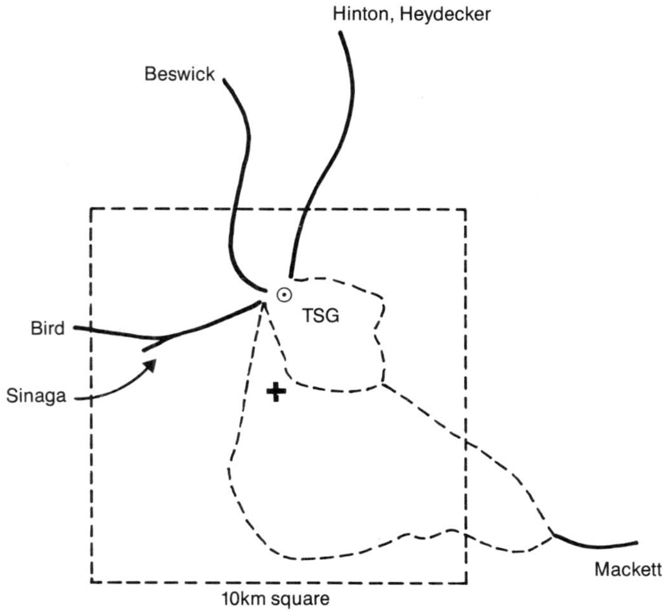

Figure 3.4 Rail journeys in central and inner London (1987)

from a series of surveys carried out by some of my colleagues on their journeys to the Transport Studies Group (TSG) at University College London during the year 1987 and reported in Mogridge (1988b).

The TSG is situated within walking distance of three main BR termini, Euston, St Pancras and King's Cross, and also within walking distance of Euston Square, Warren Street and Russell Square London Underground stations, giving access to the Victoria, Northern and Northern City, Piccadilly, Metropolitan and Circle Lines. All the other BR termini, and the Central, Bakerloo, Jubilee and District Underground lines are thus accessible via one interchange.

The journeys may be divided into those from inner London, mainly by Underground, and those from outer London and beyond by British Rail suburban services. We will discuss the former here, and examine the latter in the next chapter.

Each journey is door-to-door. One journey has alternative routes, depending on the first train to arrive at the 'home' station. The journeys are given in the order of the direct distance from home to TSG and are shown in Figure 3.4. As for car journey times given earlier in Figure 2.7, we plot the cumulative distributions of journey times on logarithmic probability scales, which would give straight lines for the distributions if they were logarithmically normal; these are given in Figures 3.5.

We then take the median journey times from the graphs, and the direct distances, and calculate median direct journey speeds. The coefficients of variation of the logarithms of journey time can be obtained directly from the slopes of the cumulative distributions. The total access times are also noted. These results are given in Table 3.12.

The journeys are as follows, in order of direct distance.

1) Sinaga. *Paddington-Euston Square*. A total of 27 journeys in June–July was recorded. This line, the Circle, has problems in the

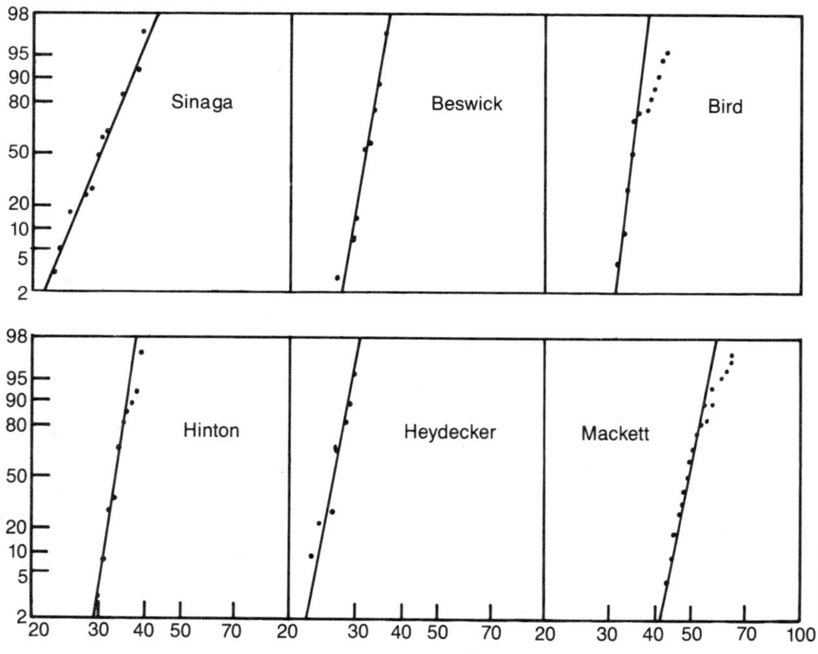

Figure 3.5 Cumulative distribution of rail journey times

Table 3.12 Rail journeys in central and inner London (1987)

	direct dist: miles	*median time mins.*	*direct speed mi/h*	*coeff: of vari:*	*access time mins.*	*percent in access*
Sinaga	2.19	30.2	4.34	0.18	16	53
Beswick	3.25	31.9	6.08	0.07	8	25
Bird	3.31	34.0	5.83	0.05	16	47
Hinton	4.0	33.4	7.26	0.06	18	52
Heydecker	4.5	26.5	10.11	0.09	12	43
Mackett	7.6	48.3	9.43	0.09	14	29

morning peak due to merging with the Hammersmith and City branch of the Metropolitan at Edgware Road, and the rest of the Metropolitan at Baker Street. A high coefficient of variation of journey time was therefore found – 0.18 – though this is still comparable to that for equivalent kerb-to-kerb car journeys.

This journey would have an estimated average speed of 3.86 miles/h from the equation in Table 3.10, compared with the recorded median speed of 4.34 miles/h.

2) Beswick. *Highgate-Euston.* A total of 29 journeys over the period January-July were recorded (by a part-time colleague). The Barnet Branch of the Northern line merges with the Edgware branch at Camden Town; nevertheless, a low coefficient of variation was found, 0.07. This journey, like all the others except the next one, has higher speed than would be estimated for the distance within the 10-km square, because they comprise journeys where the station spacing is higher than the station spacing in that area.

3) Bird. *Westbourne Park-Euston Square.* A total of 56 journeys were recorded over April-July. Although this journey has the same merge problems as that of Sinaga, this is more evident in the high tail. If we take just the lower 75 percent of the distribution, the coefficient of variation is only 0.05, but one in four journeys make the overall coefficient rather higher. This journey, where Westbourne Park is right on the edge of the 10-km square, has an estimated average journey speed of 4.86 miles/h compared with a median journey speed of 5.83 miles/h.

4) and 5) Hinton and Heydecker. *Turnpike Lane-King's Cross.* These Piccadilly line journeys start from different homes. Hinton

recorded only 26 journeys in the peak over the year 1987, but with a low coefficient of variation of 0.06 ignoring the top tail, and 0.08 with. Heydecker only recorded 15 journeys in the peak over the year, as he usually travelled by bicycle; his coefficient of variation was a little higher at 0.09. If we only take the rail part of the journeys, Hinton had a median time of 15.9 minutes and Heydecker 15.0 minutes, which given the sample sizes is comparable.

Interestingly, Heydecker's times by bicycle (median 24.1 minutes) were less than his times by rail (26.5 minutes), giving a median direct speed of 17.9 km/h (11.1 miles/h). This is probably faster than most people could manage, but is faster than car in inner London as we saw earlier.

6) Mackett. *Blackheath-Euston*. Mackett travelled via London Bridge and the Northern City, Charing Cross and the Northern, or Victoria and the Victoria line, depending on which train arrived first. There were 164 journeys recorded over the year in the peak. The coefficient of variation is 0.09 over the lower 75 percent of the distribution; including the rest gives about 0.15.

It can be seen that direct speed increases with distance, as one might expect. This is partly due to the rail stations becoming further apart as the distance from the centre increases. The proportions of the total journey involved in access do not vary systematically with distance in this small sample, but it can be seen that they comprise from 25 to over 50 percent of the journeys.

Where the journey involves a single stage, the coefficient of variation is low; as the journey becomes more complex or there is merging of lines involved, the coefficient of variation rises. These coefficients do not, however, rise above those experienced on central London car journeys.

It is not possible to give a running speed of the trains from these data, as the access time only includes to and from the station; time walking inside the station and waiting is thus included together with running time.

The GLC Development Plan Report of Studies (1968) quotes, however, a rail running speed of 15 miles/h for the lower end of the range of speeds; that is, appropriate for the central area (para. 6.37). Using a route factor of 1.28, this corresponds to a direct running speed of 11.7 miles/h, which is very close to the estimated value (12.3 miles/h) in the 1965 central area survey.

As for the bus services, we can take the rail speeds right back to the beginning of the century in the Report of the Royal Commission of London Traffic (1906), which I noted in Mogridge (1988a) and the Department of Transport has subsequently republished (DTp, 1989b).

For the underground trains, we have two sets of speeds in the Commission's report, that in Volume 3, Appendix 6, Table 44, (i) and Volume 7 Table 18 (ii).

Table 3.13 Running speeds of underground trains (1904)

	Speeds miles/h		*Distance miles*
	(i)	*(ii)*	
Steam rail			
Inner Circle	11.1	10.72	13.00
Metropolitan & District			
(Ealing to Whitechapel)	13.0 11.9		12.74
branches		14.15	
Metropolitan branches		17.21	
Electric tubes			
City & South London	14.1	13.30	6.11
Waterloo & City	22.5	18.00	1.5
Central	14.2	13.80	5.93
Great Northern & City		16.15	3.5

It can be seen that the steam Inner Circle is rather slower than the electric tubes, though as the spacing between stations increases on the Metropolitan and District branches the running speed increases. The Waterloo and City is faster as it is a non-stop service for 1.5 miles, whereas the other lines include stops at stations. The Central has 11 stations in 5.93 miles, and the City and South London also 11 stations in 6.11 miles, or roughly one station every 0.6 mile. The Great Northern and City, which ran in 1904 from Finsbury Park to

Table 3.14 Running speeds of suburban trains (1904)

	no of trains	*speed (miles/h)*
Paddington	1	21.8
Euston	5	26.9
St Pancras	4	23.5
Kings Cross	14	22.7
Moorgate (M)	6	15.0
Moorgate (GN)	6	15.7
Moorgate (SEC via Snow Hill)	19	15.7
Broad St	29	18.0
Liverpool St	34	19.6
Fenchurch St (GN)	5	21.9
Fenchurch St (GE)	13	16.0
London Bridge (LBSC)	23	18.8
London Bridge (SEC)	4	27.1
Cannon St	8	17.6
Holborn Viaduct	16	17.9
Charing Cross	12	16.5
Waterloo	27	20.9
Victoria (LBSC)	10	16.8
Victoria (SEC)	8	16.6

Moorgate, comprised six stations in 3.5 miles, or one station every 0.7 miles.

It may be noted that the Commission was of the opinion, and remember this is only 16 years after the first electric railway was opened, that:

> 'So long as trains are hauled by steam engines, we do not think that much improvement in the speed of suburban train services can be looked for. When electric traction is adopted, the advantage of quicker acceleration will be secured . . . and will enable . . . the journey times of trains to be reduced substantially'.
>
> Report volume 1 paragraph 156
> Royal Commission 1906

It can be seen that the running speeds of the central London electric tube system is already established at about 15 miles/h in 1904, the same speed as today. In fact, in 1890, when the City and South London tube was opened between King William Street and Stockwell, the design speed was stated to be 15 miles/h which, as we can see, it did not quite achieve at the beginning.

The final set of speeds for 1904 in the Commission's report is for suburban train services. These are given in two forms. In Volume 3, Appendix 6, Table 45 they are a set of speeds of the average service into the terminals during the morning rush hour, (i.e. 8–9 am.) as shown in Table 3.14.

It can be seen that these speeds are all above 15 miles/h, and rise to 27 miles/h for the longer services of the South East and Chatham into London Bridge.

A somewhat different way of looking at speeds was presented in Volume 7, Table 20, where particular stations were selected and speeds into the central London terminus for slow and fast services found. This gave the following set of speeds.

Table 3.15 Running speeds of suburban trains (1904)

	speed (miles/h)	
	slow	*fast*
Finchley–Broad Street (North London line)	15.00	19.02
Enfield–Liverpool Street	15.50	23.90
Acton–Bank (inc. tram)	8.25	10.53
East Ham–Fenchurch Street	13.85	22.50
Carshalton–London Bridge	18.75	32.00
Lower Tooting–Bank (inc. tram)	10.50	
Lower Tooting–Westminster Bridge (tram)	7.89	
Raynes Park–Bank (via Waterloo)	15.41	19.03
Richmond–Bank (via Waterloo)	15.00	22.80
Richmond–Mansion House (Underground)	12.35	

Interestingly, the Advisory Board of Engineers then added on a 10-minute walk at each end of the journey to represent access, the first time that door-to-door journeys had been estimated.

It can be seen that these running speeds are somewhat lower than those in the previous table, because these journeys include interchanges, and some include on-street tram journeys. It was suggested that a journey of 10 miles into the centre would take an average time of 40 minutes, at a speed of 15 miles/h; with access, this would mean a total journey time of one hour, or a speed of 10 miles/h. Most of these journeys would have been by steam trains which, as already noted, were expected to be speeded up when the railways were electrified.

Thus what we have so far is a tentative conclusion that running speeds have been constant in central London (strictly the City) since the beginning of this century and over a wider central area since at least 1936. Access times will probably not have changed much

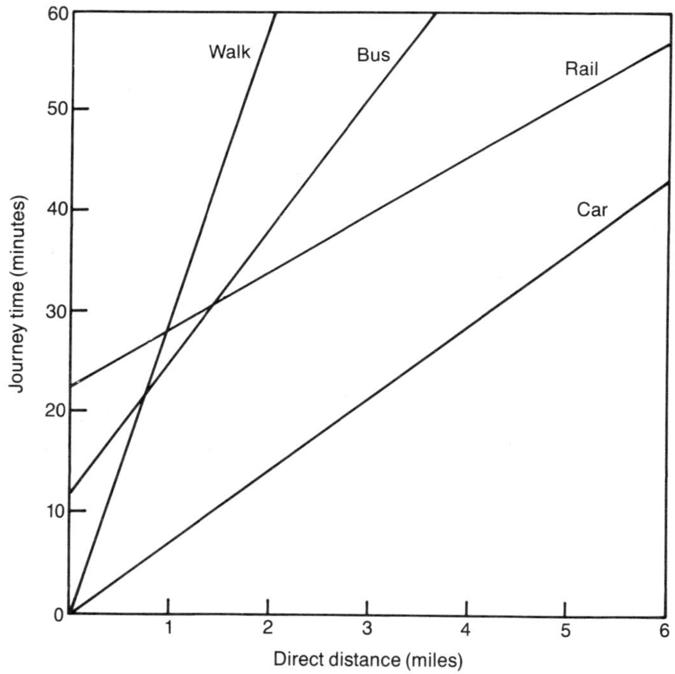

Figure 3.6 Journey time/direct distance relationships

either, since there has been little change in the actual networks since the establishment of the rail system.

The relationships from the regression analyses can be used to obtain a journey time/direct distance diagram showing which mode is quicker for any distance. This is given in Figure 3.6.

For car, Table 2.1 gives a running speed of 8.3 miles/h in 1988 but no access time. Estimates of this speed have varied slightly over the previous 20 years from 10.2 (1962), 10.7 (1969), 11.2 (1972), 9.6 (1977) to 9.2 (1983). For buses, Table 3.2 gives an intercept access time of 11.9 minutes and a direct speed of 4.5 miles/h in 1988; the latter value has varied over the years from 5.7 (1962), 5.3 (1972), 4.5 (1977) to 5.5 (1983). Table 3.8 for rail gives an intercept access time of 23.6 minutes and a direct speed of 11.1 miles/h in 1983 and an intercept access time of 22.4 minutes and a direct running speed of 10.3 miles/h in 1988. We can estimate also a walk direct speed of two miles/h.

The most intriguing point about this is that the rail and car direct running speeds are fairly similar and within the sample errors of the surveys.

Whilst bus is quicker than rail up to about 1.4 miles direct distance, walk is quicker than bus up to about 0.7 miles.

The question is, then: is there any way we can estimate car access times in the central area?

The first surveys of parking and access times in central London seem to have been made as part of the study of the potential impact of introducing parking meters to control parking in 1958, and subsequently to study the impact of raising charges, in one case by quadrupling charges. These studies are first reported by Smeed and Wardrop (1964), but are given in full in Inwood (1966) and cover the period 1954–65.

Some 31 specific addresses were visited, though not all the same ones in each survey. These were by no means random addresses; they were chosen precisely because they were well known; for example, the National Gallery. The procedure was to drive to the address required, and then search for a parking place – and later a metered parking place. The time taken to search for a parking place and park, and the time taken to walk to and from the address required, were noted.

The average time to park and access the address was often over 10 minutes just before charges were raised, and seldom below five even after a quadrupling of charges.

Unfortunately, these surveys were not continued by the GLC after 1965.

Smeed and Wardrop, basing their assumptions on these results, chose to assume values of access time of five and 10 minutes in their calculations of car journey times. Ten minutes is also the time assumed by Webster and Oldfield (1972, Table 2).

There is a further set of results available in the 1984 surveys by May and Turvey (1986), which were done as part of the analysis of the potential impact of wheel-clamping and tow-away on car parking in the central area. These confirm that average search and walking times are well over 10 minutes in congested areas and about five even in less congested areas.

It would seem to be at least feasible, therefore, that parking and access times for cars, adding both ends of the journey together, could give an intercept time on the regression equation equal to that of rail for the same direct running speed.

4 Journey Speeds: Actual Journeys

So far, we have been examining potential random journeys in the 10-km square. We now move on to examine actual journeys.

When in 1983 I became involved in the random surveys of journeys in central London, I had already had considerable experience of using the Greater London Transportation Surveys (GLTS), starting in 1966 with the proof reading of the Volume 2 report of the London Traffic Survey to familiarize me with what had been achieved up to then. It was therefore not surprising that I should wish to see whether the household data of actual journeys as sampled by the GLTS could be used to shed any light on the speeds of journeys within the 10-km square.

I had just completed with a colleague (Mogridge and Holden, 1983) a study of the spatial variation in average daily travel times in London, using the GLTS of 1962 and 1971. I had earlier (Mogridge, 1980a) completed a study of the direct journey speeds of buses and cars in London using just the 1971 survey.

We were, therefore, all set up and ready for such a comparison between actual journeys in the 10-km square and the random journeys. The 1981 GLTS was, however, at this time still being processed, so could not be included in our preliminary investigation.

There had been, however, a considerable amount of work required before we were in such a position.

The 1971 survey, undertaken by the GLC itself through a specially-created unit, is a standard household survey on an electoral register basis and covered approximately 50,000 households in an area somewhat larger than the GLC area, to cover the main contiguous built-up area.

It should be noted, however, that about 15 percent of destinations inside the area come from origins outside the area which are not covered in the household survey.

The final sample was about 40,000 households with complete records—though some of this involves 'patching', ie imputing values, particularly of household income, to records for which values were not given. The sampling rate is thus about 2 percent, though some areas have a higher sampling rate. I describe the elaborate grossing-up process below.

The 1962 survey, undertaken before the formation of the GLC in 1965, has a number of features which substantially complicate comparison with the later surveys. It was undertaken by consultants, and as the first, major, large-scale UK study, was based on US experience. It should be noted that the very first household survey in London, the London Travel Survey of 1949 (London Transport Executive, 1950), had only 3,000 households and the subsequent 1954 survey 4,000 (London Transport Executive, 1956). These earlier surveys will be mentioned again later.

The controlling body of the 1962 survey was an amalgam of national and local government bodies and the major transport operators. The survey was initially conceived as a car-owning survey only, with a sample frame drawn from the licence records of the appropriate licensing authorities. This gave problems in locating cars, since licences were given not by car body type but by type of ownership, (for example, private). This gave perhaps a 4 percent error problem. The second difficulty was to locate cars used by households as opposed to those in the hands of dealers, fleet cars, hire cars etc (possibly a 10 percent error problem in 1962). A final problem was that of unlicensed cars used illegally, about which nothing was known—though later estimates put the number of unlicensed cars in the 1970s as high as 9 percent.

This procedure will naturally give households a probability of being selected proportional to the number of cars they have. There was about 4 percent multi-car ownership in 1962.

The description of the survey procedures in the London Traffic Survey Volume 1 (1964) is, unfortunately, ambiguous on this point. We have now realized that the tapes actually contain all households with complete records, with the multi-car-owning households given different grossing-up factors. The results presented here have the same grossing-up factors for each level of car ownership and to that slight extent will be incorrect.

Later a non-car-owning household sample was added in, with all car-owning households picked up in this electoral register sample frame not being interviewed; crucially, the number of such households in each sampling district was recorded, though not their car ownership level. The two sampling rates, for car-owning and non-car-owning households, are very different, the former being about 3.5 percent and the latter 0.5 percent.

The original tapes were grossed-up against provisional census 1961 population values (number of households per sampling district) using the non-car-owning sample values of car-owning household sample numbers per sampling district as a control for the car-owning household sample frame.

The 1971 household survey grossing-up procedure was more complex, because it was realized that the multi-worker, multi-member households tended to have much lower completion rates than those of smaller households, with the exception of one-worker, one-person households which were also very difficult to contact. For each GLC borough, different grossing-up factors were applied to each household category, to ensure that the correct number of households of each worker and person structure was achieved at borough level. The process is described in Stroud (1974).

During my second period at the GLC from 1973 to 1978, I had realized that to make any sensible comparisons between the 1971 survey and the earlier 1962 survey, the two had to be made as comparable as possible. The Department of Transport agreed, and supported work after I left the GLC to repeat Stroud's process on the earlier data set. This is reported in Mogridge and Bates (1980).

The problem is complex because the 1962 sampling districts are smaller than the later GLC boroughs, though in general they can be added together to form good approximations. This small size sometimes means that they do not have a sufficient breakdown of household categories at the 10 percent level given. Values have to be imputed here from neighbouring disricts.

The special problem is that the two sampling frames may well have different response rates for the different household categories, since the questionnaires and the purpose of the questions were different. Such a difference has to be ignored, since no information is available to illuminate it.

The 1961 census did not record car ownership, so the sample values found in the non-car-owning sample are the best estimate we have of car ownership levels. The final 1961 census population values are up to 5 percent different from the provisional.

For each household record on the tape, and for each trip end, there is a zonal location. Roughly 1,000 zones were defined in each case, though the boundaries are in general different between 1962 and 1971; the 1981 survey kept the same zones as the 1971 survey. Zone sizes vary from a few hundred metres across in the central area to several kilometres across in the periphery of the study area. This is to ensure a roughly comparable amount of traffic starts or finishes in each zone.

Each zone has a defined centroid, with geocodes, which represents the 'traffic centre' of the zone, ie it is not the geographic centre but the centre where the traffic begins or ends.

The 1962 set of geocodes was derived from a match between those measured by Blumenfeld during his PhD thesis at the TSG (Blumenfeld, 1972), and those derived by the GLC; thus hopefully eliminating any further errors. These matched geocodes were incorporated in the 1962 data tapes. The 1971 data tapes already had a set of geocodes for the zonal centroids, following the census initiation of geo-coding in the 1966 10 percent sample census.

We also have, for each trip record, its duration in minutes. This is a much more complex record than it seems at first sight because the time given is a perceived time by the respondent, not a time actually measured with a stop watch. There is, however, a constraint in that the time of departure and the time of arrival are both required, and these help to delimit the time actually given by the respondent.

In the 1962 survey, there is a further coding problem, in that the results are only coded as origin time and destination time to 10-minute intervals. We have to assume that trips begin and end in the centres of these intervals, which may well lead to substantial bias for certain kinds of trip time distributions. In the 1971 survey, trip times are coded both this way and more directly, so that a comparison can be made between the two methods of recording.

Even if times of trips were recorded in one-minute intervals, there would still be the problem that people think, and therefore record, their trip times in multiples of five minutes up to 30 minutes; and then in multiples of 15 minutes. Analysis of the GLTS 1981 data by Beardwood (1989) shows this clearly (Figure 4.1). The proportion recording times in five-minute intervals is 79 percent.

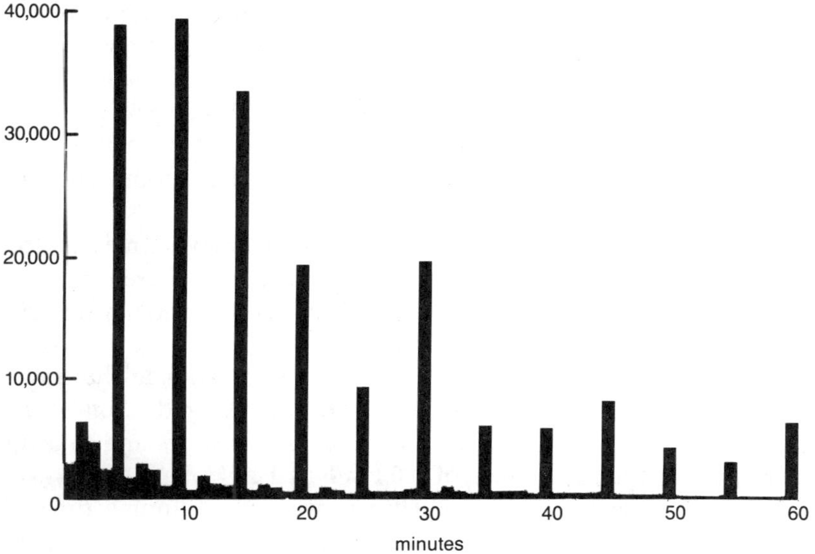

Figure 4.1 Trip time distribution GLTS 1981

This clearly limits the kind of analysis that can be done, as we will see later.

Such work as I am presenting here can be regarded, therefore, only as tentative until much more work has been done on people's perception of time and how it relates to reality.

In 1983, therefore, we started examining the 1971 GLTS data of trips in the 10-km central square by residents of the GLTS area who live in households, ie excluding non-household residents and excluding people who live outside the GLTS area who make trips inside the 10-km square. These latter categories of exclusions are not thought likely to affect the results appreciably, because their trips are expected to be similar to the trips included.

There are 3,890 car journeys whose times are known where both ends of the journey are inside the 10-km square. This is a massive sample and can be expected to give a fairly accurate result, subject to the provisos noted above.

The mean direct journey speed is 7.8 km/h (4.8 miles/h), with a mean time of 19.1 minutes and a direct distance of 2.5 km (1.55

miles). If, therefore, kerb-to-kerb direct speeds are indeed about 10 miles/h (16 km/h) as shown in Table 2.1, then a journey of 2.5 km will take 9.4 minutes, leaving the rest of the journey time of 9.7 minutes to parking and access.

It will be noted that the actual journey direct distance of 2.5 km is less than half the random journey direct distance of 5.2 km.

The comparison can be continued with the data on bus and rail trips.

There are 4,870 bus trips recorded with known times in the 10-km square in 1971. The average direct journey speed was 4.5 km/h (2.8 miles/h) for a mean time of 35.9 minutes and a mean trip direct length of 2.7 km (1.7 miles).

If we take the 1983 regression equations to estimate the travel time for a 2.7 km journey from Table 3.3, using the total time against distance we find a travel time of 32.2 minutes and thus an average direct journey speed of 5.0 km/h (3.1 miles/h).

If we use the component equations, we find a running time of 14.2 minutes, a walk of 12.2 minutes and a wait of 6.4 minutes, making a total of 32.8 minutes. This gives a running speed of 11.4 km/h (7.1 miles/h) and an average direct speed of 4.9 km/h (3.1 miles/h).

The inconsistency arises from the fact that the component equations use the one trip entirely on foot, whereas the total excludes it. It can be seen that including the walk trip increases the travel time estimate.

The fact that the random journeys are faster than the actual journeys can be explained by the fact that actual journeys would not include journeys which are quicker by another mode. Thus, although long bus trips are faster than short, long rail trips are even faster, so rail would be the preferred mode. Moreover, many of the actual bus journeys will be in the centre where road speeds are lower, rather than spread throughout the 10-km square. With these qualifications, the two surveys are in good agreement.

There are 2,595 LT rail (main mode) and 221 BR rail (main mode) journeys entirely within the 10-km square. A main mode is the mode which has the longest length. The LT journeys have an average direct length of 4.1 km (2.5 miles) taking 34.4 minutes at an average direct speed of 7.2 km/h (4.5 miles/h). The BR journeys have an average direct length of 4.4 km (2.7 miles), taking

36.3 minutes at an average direct speed of 7.4 km/h (4.6 miles/h). The LT values are thus a good approximation to the rail average.

If we take the 1983 regression equations to estimate the travel time for a 4.1 km journey from Table 3.10, using the total time against distance we find a travel time of 36.0 minutes. This gives an average direct speed of 6.8 km/h (4.2 miles/h).

If we use the component equations, we find a running time of 11.9 minutes, a walk of 18.1 minutes and a wait of 6.3 minutes, making a total of 36.3 minutes. This gives a running speed of 20.7 km/h (12.8 miles/h) and an average direct speed of 6.8 km/h (4.2 miles/h).

Thus actual rail journeys are slightly faster than the random journeys. The random journeys, however, included four journeys out of the 24 where the BR service was timetabled at a 15-minute interval or greater (especially on the North and South London lines which are orbital lines). Actual passengers would be expected to know the timetable, whereas the random journeys could not always take this into account. The difference is, however, small.

It is perhaps remarkable that a random tour of only 25 sections travelled by rail should give results so close to that of a survey comprising nearly 3,000 records; or that a household survey based on people's perception of time should be so close to a stop-watch record of time. Nevertheless, the results for the rail and bus surveys gave us confidence that we could trust the results of the household car journey survey, for which in the random survey we have no records of access times.

It will be noted in these results that the actual car trips average direct speed is 7.8 km/h whereas that of the rail trips is 7.2 km/h, though admittedly the rail trips are longer. It is this similarity, noted in my lecture (Mogridge, 1986a), that set me thinking.

The first step was to expand the analysis from the 10-km square to the whole city. To do this, we needed to work in polar co-ordinates, ie to define a centre of the city and to define all locations by distance from the centre (the radial distance). My first analysis of London using polar co-ordinates was in Mogridge (1968, 1969), when I examined the journey-to-work from the 1962 GLTS.

A common centre of the city for the surveys based on the 1966 census spatial distribution of population in the South East region out to 35 kilometres (i.e. beyond GLC boundaries) was defined by

Figure 4.2 The central 6-km radius circle

Mogridge (1973). These census data define population location by 3,500 wards in the region, each ward with a geo-code.

This centre is nearer Piccadilly Circus than Charing Cross, the traditional grid centre used for the 10-km square grid. Its grid reference is 529510 180680. The movement of this centre of population through time since 1801 can be followed in Mogridge (1985). It is drifting slowly westwards through time at the rate of a few metres per year, so that this movement can be ignored.

Thus the cover design for this book has been chosen to incorporate my photograph of a traffic jam at Piccadilly Circus, superimposed on a local street map of the area.

The central area in a polar co-ordinate representation thus needs to be a circular area. We chose a circle of radius 6-km as the defining boundary, both because the area within the circle (113 sq.km) is approximately the same as the 10-km square (100 sq.km) and, more particularly, because we wanted to ensure that we had a big enough sample of car journeys in the analysis.

The area of the central conurbation is in fact only 27.3 sq.km, equivalent to a 3-km radius circle, but the number of car journeys recorded as terminating in this area in the surveys is too small for one to have any confidence in the results. Choosing a larger centre has other problems, however (as we shall see), the most important of

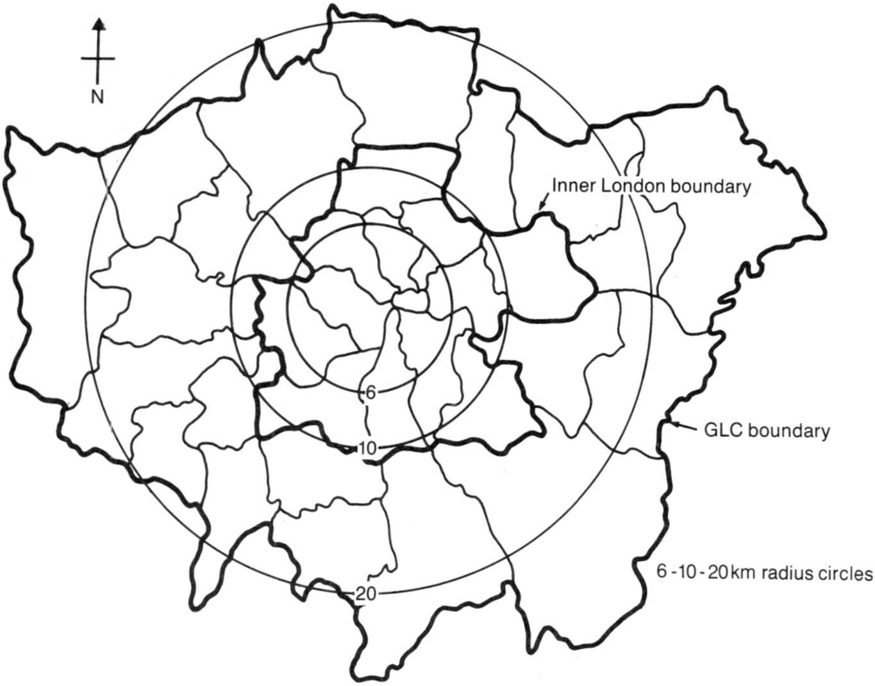

Figure 4.3 Circular bands around centre of London with administrative boundaries

which is that journeys do not tend to penetrate the centre very far. Putting this another way, people tend to live on the same side of the centre as they work, in order to avoid having to travel across it.

The 6-km radius circle is shown in Figure 4.2.

With this circle as a basis we can now repeat the above analyses for car and rail journeys, ignoring bus journeys for the moment. We take the analyses right out to the edge of the conurbation for trip origins, dividing the area up into 2-km wide bands around the centre. This is shown in Figure 4.3.

We start with car trips in 1971.

Table 4.1 Characteristics of car journeys to the centre (GLTS, 1971). Main mode journeys, destination within 6-km of centre, 24 hour

Origin ring (km)	*N*	*T (mins)*	*L (km)*	*S (km/h)*
0 – 2	72405	22.72	2.86	7.54
2 – 4	145417	17.80	2.42	8.17
4 – 6	183584	17.18	2.57	8.97
6 – 8	99931	22.15	4.41	11.92
8 – 10	50462	29.05	6.82	14.09
10 – 12	42549	35.99	8.98	14.96
12 – 14	34946	42.93	11.11	15.51
14 – 16	30056	45.50	13.51	17.82
16 – 18	26811	49.16	15.43	18.83
18+	55212	62.06	26.35	25.48
Total	741373			

Residents of GLTS area only
N = grossed-up trip numbers. T = average trip time. L = average direct distance.
S = average direct speed

The following points can be made about these data. First, it can be seen that the speed already measured for journeys with both origins and destinations in the 10-km square (7.8 km/h) is about the same as the speeds given here for origins in the three central rings and destinations within the 6-km circle.

Secondly, it can be seen that the total number of trips per day with a destination within the 6-km circle is estimated at 741,373. Doubling that for total travel, we see that about 1.5 million car trips are made every day to or from the centre. This is in approximate agreement with the number of trips estimated to cross the central area cordon each day, if one makes allowances for through trips, trips which begin outside the GLTS area and trips which are made by non-household residents.

The total number of trips with origins within the 6-km circle as well as destinations is 401,406, or 54 percent of the total. This demonstrates the point about just how short most car trips are. We will come back to this point in Chapter Nine.

We can see that the number of car trips tails off rapidly as the distance of the origin ring from the centre increases, remembering that we do not have in these data trips which begin outside the GLTS, or approximately 20-km from the centre in some directions.

As the origin distance from the centre increases, the average direct speed of the journey goes up, reaching 25 km/h for the longest trips over 18 km with an average trip length of 26 km. It can be seen, however, that travel times do not rise consistently, but fall slightly from the innermost ring.

The most intriguing point, however, is that the average trip length, apart from that of the trips in the 6-km circle, falls some way short of the distance necessary for the average trip to reach the centre. However, as the trip length increases, the penetration of the trip increases. For the 6–8 km origins, the trip length of 4.41 km has a shortfall of 2.59 km from the centre, assuming the average trip begins at the mid-point of the ring. For the 16–18 km origins, the trip length of 15.43 km has a shortfall of only 1.57 km from the centre. This one would expect as the access time becomes a lower proportion of the total time of the trip.

If we examine the marginal speed of the trips as the distance of the origin from the centre increases, we should obtain the local speed of travel. Table 4.2 gives the calculations of the increase in travel time for each increase in travel distance.

Now if all origins were centred in each ring and all destinations were in the centre of the city, then as we move outwards from ring

Table 4.2 Local speeds of car trips

Origin distance (km)	ΔT *(mins)*	ΔL *(km)*	ΔS *(km/h)*
6	4.97	1.84	22.21
8	6.90	2.41	20.96
10	6.94	2.16	18.67
12	6.96	2.13	18.76
14	2.57	2.40	56.03
16	3.66	1.92	31.48
18+	12.90	10.92	50.79

origin distance is boundary between two adjacent rings.
ΔT is difference in average trip time between two rings. ΔL is difference in average direct trip length. ΔS is marginal direct speed.

to ring, the trip length would increase by two km for each additional ring. We can see, however, that trip lengths as we move outwards increase by rather more than this in general; implying, as already noted above, that trips are penetrating more deeply into the centre as the origin distance increases. This probably means that access time is increasing, as access is more time-consuming the nearer the centre.

The marginal direct travel speed, therefore, will usually be an underestimate of the true local direct speed of travel. We can see that, whilst the speeds increase as the distance of the origin from the centre increases, the increase is erratic, so that we cannot draw any firm conclusion from these results; apart from the fact that the local speed is greater than the average speed, as it must be if average speed is increasing with distance from the centre.

The actual road speed will be greater than the direct speed by the route factor, which one can expect to be about 1.2 for trips to the centre. At a distance of about 20 km from the centre, a 30 km/h direct speed therefore corresponds to an actual road speed of about 36 km/h (about 20 miles/h), which as we have seen in Table 2.4 is about the average all-day speed in the outer area of London. The results are therefore consistent even if not defined very accurately.

It is also probable that there are some sample errors in the results—ie miscoding—since the results are so erratic. We shall investigate this further shortly.

We now move on to rail trips. First of all, we will keep the two systems, BR and LT, separate. As for car, we give the 24-hour trips into the 6-km circle from each origin ring from the 1971 GLTS.

Table 4.3 Characteristics of rail journeys to the centre (GLTS, 1971). Main mode journeys, destination within 6-km of centre, 24 hour

	LT				*BR*			
Origin ring (km)	*N*	*T (mins)*	*L (km)*	*S (km/h)*	*N*	*T (mins)*	*L (km)*	*S (km/h)*
0 – 2	95909	32.68	3.85	7.07	4185	35.68	4.47	7.51
2 – 4	92804	32.33	4.11	7.62	5145	36.76	4.38	7.14
4 – 6	110243	36.03	4.84	8.05	11251	33.48	4.49	8.04
6 – 8	75943	40.83	6.72	9.87	23324	38.48	6.19	9.65
8 – 10	43709	47.24	8.31	10.56	25792	46.65	8.51	10.95
10 – 12	54362	49.51	10.05	12.18	36999	51.42	10.34	12.07
12 – 14	48331	53.07	12.12	13.70	36716	51.15	12.23	14.35
14 – 16	32657	57.14	14.35	15.06	37395	56.46	14.45	15.36
16 – 18	31353	59.54	16.09	16.21	46659	58.39	16.29	16.74
18+	40954	65.60	20.50	18.75	120507	70.76	26.31	22.31
Total	626265				347973			

Residents of GLTS area only.
N = grossed-up trip numbers. T = average trip time. L = average direct distance.
S = average direct speed

The first thing to note about this table is the similarity between the LT and BR speeds. The BR speeds in the outermost area (ie beyond 18 km) are greater because the trips come from further out in the GLTS area, on average 26 as opposed to 20 km. Apart from this, the differences in speed are negligible.

This is all the more remarkable because the two systems are run in quite different manners owing to the history of their development. BR was essentially a long-distance, steam railway in conception, with terminals on the edge of the central area; whereas LT was *ab initio* an electric railway that ran right through the centre. The signalling and control, and the frequencies of service, are different. Nevertheless, the two trip speeds are very similar.

It will also be noted that the maximum number of LT trips begin in an inner ring, between four and six km, and tail off as we proceed outwards; whereas BR trips continue to increase as we proceed outwards. In fact many BR trips begin outside the GLTS area, and therefore will not be recorded here.

It will also be noted that these speeds are a little lower than the speeds for car given in Table 4.1, but the rate of increase with distance from the centre is comparable.

One factor that may be relevant to this is that rail trips are longer for a given origin ring, ie they penetrate further into the centre. The difference for the longer trips is about one km; for the shorter, the difference is nearer two km.

If we examine the marginal or local speeds as for car, we find the results shown in Table 4.4.

These local speeds are rather more erratic than those for cars, with an actual decrease in average trip time for BR rail trips over the

Table 4.4 Local speeds of rail trips

	LT			BR		
Origin distance (km)	ΔT *(mins)*	ΔL *(km)*	ΔS *(km/h)*	ΔT *(mins)*	ΔL *(km)*	ΔS *(km/h)*
6	4.80	1.88	23.5	5.00	1.70	20.4
8	6.14	1.59	15.5	8.17	2.32	17.0
10	2.27	1.74	46.0	4.77	1.83	23.0
12	3.56	2.07	34.9	−0.27	1.89	—
14	4.07	2.23	32.9	5.31	2.22	25.1
16	2.40	1.74	43.5	1.93	1.84	57.2
18+	6.06	4.41	43.7	12.37	10.02	48.6

origin distance is boundary between two adjacent rings.
ΔT is difference in average trip time between two rings. ΔL is difference in average direct trip length. ΔS is marginal direct speed

12 km boundary giving a negative local speed! In general, however, it can be seen that local speeds increase with increasing distance from the centre.

As opposed to car trips, we might expect that rail trips would have an increasing amount of time devoted to local access in the origin ring and a much smaller difference in access time in the destination circle, since the penetration is approximately the same for all origin rings; that is, the trip length increases by two km per ring as we move outwards, unlike car trips. This we will examine later.

So what we have shown so far is that not only is the average direct speed of car trips similar to that of rail trips in the centre, it is also similar to that of rail trips for any origin distance from the centre for trips to the centre.

Moreover, if the average direct speeds were the same, then the local direct speed of car travel at a given distance from the centre would be the same as the local direct speed of rail travel.

Since the two rail systems are so similar, we will continue the analyses by merging the two sets of rail data.

The most important feature of the rail systems is that they are predominantly used for the journey to work in the morning peak and the return journey in the evening. This is somewhat different to the car journeys which, although peaked, do have a bigger spread through the day.

It therefore seemed reasonable to examine peak-hour travel as a separate exercise, especially since if there were to be any connection between the speeds on the two systems—road and rail—they must be contemporaneous. We know already from the examination of road speeds that, while peak-hour road speeds in the centre are little different from midday off-peak, in the inner and outer areas of London this is not so, peak-hour travel being slower than off-peak.

At this point, too, we are going to introduce the analyses of 1962 and 1981 data, the latter having become available by the time we reached this stage in our analyses.

We are going to drop the analyses of the central six km origins, but extend the analyses to 24 km from the centre, although at this distance the GLTS area does not cover the whole ring. The morning peak-hour is defined as 7–10 am, and we examine all destinations in this period in the central six km circle.

We give first the numbers of the samples of trips, to demonstrate that we are dealing with a limited sample size and therefore cannot place too much confidence in the results. It must be said that sample sizes of this magnitude are in general thought to be extremely large in most survey work, we are concerned however to measure accurately quite small differences in average speeds, and for this the sample sizes are only just tolerable.

It should be remembered that the procedures for the 1962 sample were somewhat different, with a much higher sampling rate for car-owning households; this also gives a much higher sample size for rail trips, because many of the rail trips are from car-owning households. The car-owning sample size in 1962 was approximately 30,000 households, whereas if the same procedures had been used as in the later surveys it would have been approximately 15,000. It can be seen that the sample sizes of car trips rose between 1971 and 1981 and fell for rail. The precise values depend on population change as well as on car ownership changes.

In general in the period 1962 to 1981 the number of people

Table 4.5 Sample sizes in the morning peak-hour analyses of the three GLTS samples of journeys by distance of origin from centre, central area destinations between 7–10 am

origin distance (km)	*1962*		*1971*		*1981*	
	car	*rail*	*car*	*rail*	*car*	*rail*
6 – 8	607	633	319	670	374	613
8 – 10	470	729	184	511	249	553
10 – 12	458	938	195	801	235	694
12 – 14	481	1068	181	774	226	636
14 – 16	470	1355	171	684	275	760
16 – 18	334	1381	171	879	134	581
18 – 20	248	1033	117	669	112	402
20 – 22	163	671	88	574	84	322
22 – 24	127	555	71	377	36	199
Total sample	3358	8363	1497	5939	1725	4760

Table 4.6 Minimum travel speeds in the morning peak-hour analyses of the three GLTS samples of journeys by distance of origin from centre, central area destinations between 7 – 10 am

origin distance (km)	*1962 car (km/h)*	*1962 rail (km/h)*	*1971 car (km/h)*	*1971 rail (km/h)*	*1981 car (km/h)*	*1981 rail (km/h)*
6 – 8	1.36	2.09	2.00	2.56	2.35	1.47
8 – 10	3.83	4.87	3.88	3.50	4.48	2.62
10 – 12	3.92	5.33	5.36	5.15	5.24	4.01
12 – 14	5.21	6.66	6.69	2.97	4.98	5.71
14 – 16	6.02	5.62	7.51	6.92	4.95	4.66
16 – 18	8.21	7.76	8.92	6.82	7.63	7.41
18 – 20	8.51	9.46	9.41	7.76	7.00	7.61
20 – 22	10.04	9.32	11.10	9.28	9.01	8.78
22 – 24	7.76	9.93	8.00	9.70	10.26	10.24

entering central London in the morning peak fell from 1.25 million to 1 million approximately, a drop of 20 percent, although this includes through trips and trips by household members from outside the GLTS area and non-household residents. It can be seen that the data reflects this, although only partially because this is the raw data, not grossed-up.

One of the tests on the raw data is to look at the minimum and maximum speeds of travel recorded, as this will probably throw up any mis-coding. We do this in Tables 4.6 and 4.7.

It can be seen that minimum speeds for journeys from the inner rings are very low indeed, only just over one km/h in one case, though generally over two. If we examine the values quoted in the studies by Smeed and Jeffcoate and Mogridge and Fry, we see that such direct speeds are not beyond the bounds of possibility, although relatively rare. From the peripheral rings, a speed of about 10 km/h seems to be a common low, giving a two-hour journey from 20 km. Again this is possible, as a result of accidents, traffic jams, breakdowns etc.

When we look at the maxima, a different story emerges. It can be readily seen from Table 4.7 that some quite unbelievable speeds are

Table 4.7 Maximum travel speeds in the morning peak-hour analyses of the three GLTS samples of journeys by distance of origin from centre, central area destinations between 7 – 10 am

origin distance (km)	*1962*		*1971*		*1981*	
	car (km/h)	*rail (km/h)*	*car (km/h)*	*rail (km/h)*	*car (km/h)*	*rail (km/h)*
6 – 8	60.01	42.30	36.72	35.70	256.20	25.80
8 – 10	79.08	75.72	49.62	49.68	30.18	21.54
10 – 12	73.74	67.80	42.84	42.54	65.16	24.18
12 – 14	96.96	77.88	89.94	36.84	54.84	29.82
14 – 16	88.14	86.40	67.08	34.56	78.00	28.07
16 – 18	98.76	89.52	45.30	44.82	66.24	37.62
18 – 20	115.80	119.88	114.84	41.70	78.48	43.74
20 – 22	69.54	129.36	144.66	40.44	478.80	33.06
22 – 24	53.70	68.16	47.64	39.30	96.48	31.50

Table 4.8 Coefficients of variation of travel speeds in the morning peak-hour analyses of the three GLTS samples of journeys by distance of origin from centre, central area destinations between 7 – 10 am

origin distance (km)	*1962*		*1971*		*1981*	
	car	*rail*	*car*	*rail*	*car*	*rail*
6 – 8	0.468	0.398	0.421	0.326	1.075	0.339
8 – 10	0.463	0.363	0.430	0.340	0.373	0.288
10 – 12	0.486	0.323	0.375	0.283	0.426	0.242
12 – 14	0.433	0.289	0.459	0.280	0.344	0.235
14 – 16	0.455	0.281	0.394	0.217	0.424	0.231
16 – 18	0.547	0.271	0.313	0.251	0.412	0.227
18 – 20	0.433	0.315	0.534	0.227	0.459	0.224
20 – 22	0.321	0.326	0.636	0.227	1.917	0.205
22 – 24	0.315	0.258	0.327	0.210	0.602	0.213

given, including one car journey at nearly 500 km/h in 1981, and several journeys at over 100 km/h (60 miles/h).

The first step in seeing what credence can be given to these high values is to examine the coefficient of variation in the results, to obtain a feel for the probability as opposed to the feasibility. This we do in Table 4.8, calculated on the unweighted raw data.

Given the results from the coefficients of variation quoted by Smeed and Jeffcoate and Mogridge and Fry for road running speeds, we would not expect a car journey coefficient for a single journey less than 0.15. Similarly, for rail journeys, the evidence quoted above would suggest perhaps a slightly lower coefficient, but of the same order.

The problem in estimating what we might expect the coefficient for our sample to be is that in order to obtain a sufficient sample size to define the means accurately, we have chosen a fairly large central area; which therefore gives the consequence that trip lengths vary, especially for the inner rings. We also have many different journeys where access time is varying at both ends. Moreover, as we saw in Figure 4.1, times are in general only given to the nearest five-minute interval. Furthermore, zone size is increasing as the origin distance

Table 4.9 Standard errors of mean travel speeds in the morning peak-hour analyses of the three GLTS samples of journeys by distance of origin from centre, central area destinations between 7 – 10 am

origin	*1962*		*1971*		*1981*	
distance (km)	*car (km/h)*	*rail (km/h)*	*car (km/h)*	*rail (km/h)*	*car (km/h)*	*rail (km/h)*
6 – 8	0.239	0.172	0.316	0.137	0.706	0.127
8 – 10	0.342	0.169	0.527	0.185	0.322	0.131
10 – 12	0.392	0.150	0.436	0.133	0.410	0.112
12 – 14	0.352	0.136	0.595	0.154	0.369	0.130
14 – 16	0.409	0.128	0.564	0.136	0.448	0.124
16 – 18	0.631	0.130	0.468	0.149	0.676	0.154
18 – 20	0.602	0.187	1.058	0.163	0.914	0.197
20 – 22	0.568	0.259	1.528	0.196	5.510	0.214
22 – 24	0.654	0.238	0.851	0.227	2.454	0.293

Table 4.10 Analysis of direct distance, travel time and direct speed of car and rail journeys in the morning peak-hour of the three GLTS samples of journeys by distance of origin from centre, central area destinations between 7 – 10 am

origin distance (km)	car				rail			
	no	dist (km)	time (min)	speed (km/h)	no	dist (km)	time (min)	speed (km/h)
				1962				
6 – 8	41400	4.92	25.52	11.57	88280	6.40	37.74	10.17
8 – 10	29120	7.60	32.64	13.97	90600	8.73	44.78	11.70
10 – 12	24890	9.41	37.63	15.00	93330	10.36	47.24	13.16
12 – 14	23670	11.34	42.59	15.97	101490	12.33	50.07	14.78
14 – 16	22770	13.54	46.93	17.31	107650	14.35	54.24	15.87
16 – 18	15160	15.53	50.14	18.58	104710	16.29	58.00	16.85
18 – 20	12650	17.51	52.43	20.04	69420	18.46	61.04	18.15
20 – 22	7590	19.88	57.45	20.76	56670	20.43	64.74	18.93
22 – 24	5610	21.58	61.55	21.04	32630	22.26	64.85	20.60
total	182860				744780			

continued

Table 4.10—continued

origin distance (km)	*car*				*rail*			
	no	*dist (km)*	*time (min)*	*speed (km/h)*	*no*	*dist (km)*	*time (min)*	*speed (km/h)*
1971								
6 – 8	51740	4.94	24.87	11.92	113910	6.77	39.64	10.25
8 – 10	31700	7.20	29.74	14.53	84020	8.52	44.84	11.40
10 – 12	27330	9.42	38.67	14.62	112960	10.15	48.70	12.51
12 – 14	24480	11.19	43.69	15.37	109550	12.12	50.76	14.33
14 – 16	23300	13.45	47.46	17.00	90960	14.34	55.13	15.61
16 – 18	20830	15.27	50.92	17.99	106820	16.18	58.26	16.66
18 – 20	12250	17.48	56.10	18.70	69660	18.30	61.60	17.82
20 – 22	9400	19.60	58.57	20.08	53720	20.16	63.67	19.00
22 – 24	7330	21.49	63.43	20.33	39160	21.70	64.30	20.25
total	208360				780760			

continued

Table 4.10—continued

origin	*car*				*rail*			
distance (km)	*no*	*dist (km)*	*time (min)*	*speed (km/h)*	*no*	*dist (km)*	*time (min)*	*speed (km/h)*
				1981				
6 – 8	43920	5.05	27.68	10.95	73990	6.64	45.40	8.78
8 – 10	31850	7.07	35.38	11.99	70110	8.48	50.06	10.16
10 – 12	28250	9.37	43.36	12.97	84650	10.40	53.81	11.60
12 – 14	26910	11.08	45.09	14.74	79230	12.32	55.74	13.26
14 – 16	33170	13.43	51.42	15.67	91280	14.36	60.98	14.13
16 – 18	16040	15.08	54.36	16.64	72420	16.32	62.45	15.68
18 – 20	14510	17.51	57.28	18.34	51580	18.19	64.55	16.91
20 – 22	13120	19.18	61.07	18.84	49700	20.15	67.09	18.02
22 – 24	5800	21.72	63.90	20.39	32970	21.99	71.26	18.52
total	213570				605930			

GLTS residents only. no = grossed-up sample size. dist = average direct distance. time = average journey time. speed = average direct speed.

from the centre increases, so the approximation of the zonal geocode becomes less valid as a determination of journey length. These effects will all increase the coefficient of variation.

As we see from Table 4.8, most of the rail coefficients are in fact below or around 0.3, and as low as 0.20 for the longer journeys in 1981, though rising to 0.35 for the inner origins. Car journey coefficients are in general somewhat higher, up to 0.45. The 1962 values are especially higher for both car and rail, and are in that respect more suspect.

The particularly high values are no doubt spurious, and caused by the very high speed values of particular journeys. We will later exclude them from the analysis.

The standard errors of the means are given in Table 4.9, to see what reliance can be placed on individual values.

The values of car journey speed are thus typically determined to within 0.5 km/h, rising with distance of the origin from the centre as the sample size falls. The values of rail journey speed are more precisely determined, to 0.15 km/h or better, and again tend to rise with distance as sample size falls. The evidence of spurious values of speed is seen in the extremely high standard errors in some of the car values.

After all this preliminary clarification of the errors in the data, we now move on to examining the results for mean speeds for car and rail in each of the three surveys. We give the grossed-up sample size, the mean distance, time and speed of journeys by car and rail in the morning peak for the three surveys in Table 4.10.

The biggest changes that have occurred are in the numbers and ratios of car and rail trips in the period. Whilst the totals here are only for the specified ranges of origins, rather than for the totals given in the all-day tables above, it can be seen that the estimated number of car trip origins, outside the central six km and up to 24 km within the GLTS area, have risen from the 1962 value of 182,860 to 208,360 in 1971, but between then and 1981 they only rose slightly to 213,570. Rail trips on the other hand rose slightly from 744,780 in 1962 to 780,760 in 1971 before falling back substantially to 605,930 in 1981.

Given the errors in these numbers, it is probably better to state that car trips rose by about 10 percent in the first period and then were stable in the second, whilst rail trips were stable in the first period and then fell substantially by about 20 percent in the second. This can be roughly confirmed by the central area cordon-crossing flows, which include through trips, trips by external residents and trips by non-residents.

As a result, the proportions travelling by car and rail (the modal split) for each origin ring have changed over the decades. This is shown in Table 4.11.

Table 4.11 Modal split by rail in the morning peak-hour of the three GLTS samples of journeys by distance of origin from centre, central area destinations between 7 – 10 am

origin distance (km)	*modal split by rail* 1962	1971	1981
6 – 8	68	69	63
8 – 10	76	73	69
10 – 12	79	81	75
12 – 14	81	82	75
14 – 16	83	80	73
16 – 18	87	84	82
18 – 20	85	85	78
20 – 22	88	85	79
22 – 24	85	84	85

These values are just the modal split between car and rail; they omit the proportion of bus travellers. It can be seen that in the second decade, with falling rail numbers, the modal split by rail has fallen at all distances of the origin from the centre (except, apparently, for the longest trips). In the first decade on the other hand no clear picture emerges.

It will also be seen that the mean direct distance of these peak-hour trips for a given origin ring is slightly longer for rail trips than for car, confirming what we have already seen for all-day trips. Whereas the average rail trip penetrates to within one km of the centre, the average car trip only penetrates to within two km of the centre.

Table 4.12 Mean journey speed by distance from centre of origin Independent regression equations

1962	$v_c = 7.74\ (0.46) + 0.630\ (0.03)\ d$
	$v_r = 6.18\ (0.44) + 0.630\ (0.03)\ d$
1971	$v_c = 8.64\ (0.54) + 0.546\ (0.04)\ d$
	$v_r = 5.82\ (0.37) + 0.636\ (0.03)\ d$
1981	$v_c = 6.66\ (0.43) + 0.600\ (0.03)\ d$
	$v_r = 4.44\ (0.32) + 0.648\ (0.02)\ d$
where	v_c = mean direct journey speed of cars
	v_r = mean direct journey speeds of rail
	d = mean distance of origin within a ring from centre (taken to be the midpoint of the ring)

The standard errors are in brackets

Table 4.13 Mean journey speed by distance from centre of origin Regression equations with forced equality of slope coefficient

1962	$v_c = 7.82\ (0.19) + 0.628\ (0.012)\ d$
	$v_r = 6.22\ (0.19) + 0.628\ (0.012)\ d$
1971	$v_c = 7.53\ (0.20) + 0.628\ (0.012)\ d$
	$v_r = 5.93\ (0.20) + 0.628\ (0.012)\ d$
1981	$v_c = 6.34\ (0.20) + 0.628\ (0.012)\ d$
	$v_r = 4.74\ (0.19) + 0.628\ (0.012)\ d$
where	v_c = mean direct journey speed of cars
	v_r = mean direct journey speeds of rail
	d = mean distance of origin within a ring from centre (taken to be the midpoint of the ring)

The standard errors are in brackets

Turning now to the crux of the matter, the journey speed data can be further analysed by regression analysis to give the relation between journey speed and distance of the origin from the centre. We use simple least squares regression analysis to derive the linear equations shown in Table 4.12.

Given the similarity of the slopes of these lines, it is worth considering what happens if all slopes are forced to be the same. We give the results in Table 4.13.

With this forced equality of the slope coefficient, we can see that these data suggest that mean car speeds are always about 1.60 km/h faster than rail speeds. Whilst there is no change in mean speeds in the decade from 1962 to 1971, mean speeds on both car and rail have fallen by about 1.3 km/h in the following decade from 1971 to 1981.

To some extent, however, we have seen above that the faster speeds of car journeys are likely to be a spurious effect of the mis-coded data. If we 'clean' the data by excluding all speeds greater than the following:

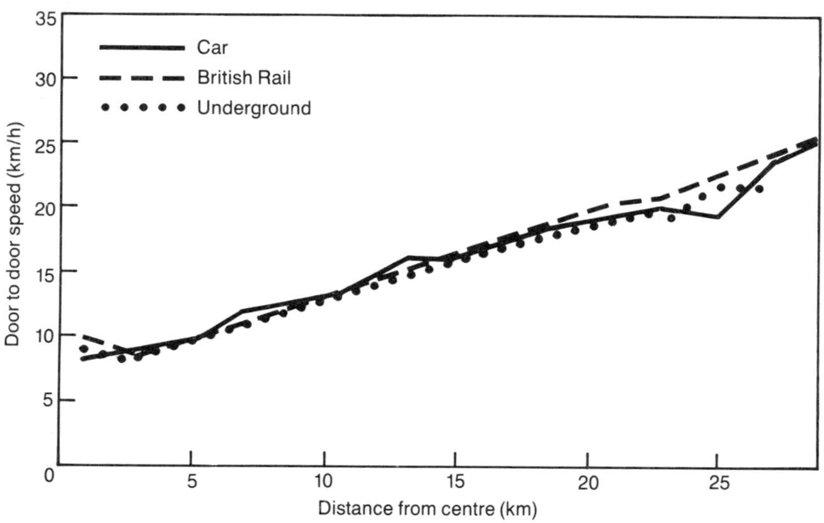

Figure 4.4 Direct journey speed by distance of origin from centre, morning peak-hour (7–10 am), central destinations within 6-km of centre, GLTS 1981

for car journeys: 80 km/h direct speed
for BR rail journeys: 50 km/h direct speed
for LT rail journeys: 40 km/h direct speed

then we find the following relation between journey speed and distance of the origin from the centre. This figure was obtained by Blase (1985) and published by Bly *et al* (1987) and is thus an independent check on our results.

It can be seen that, after this cleaning process, the direct journey speeds of car and both rail systems are very similar indeed. The difference between these results and our own results on the uncleaned data is that the high car speeds in particular have been excluded, bringing down the average direct speed of car journeys to be more comparable with the rail journeys.

What is particularly noteworthy about the equations in Table 4.13 is that when we compare the speeds of car and rail journeys with the numbers of such journeys given in Table 4.10, we can see that journey speeds were stable in the first decade with rises in car numbers and stable rail numbers, and fell in the second decade with stable car numbers and falling rail numbers.

The stability of road speeds in the decade 1962–71 has already been noted in Chapter Two, together with the fall since then. The argument about journey speeds is therefore consistent with the road speed data. To put this more bluntly, the correlation is between speed and rail flows; there is no correlation between speed and car flows.

This is a rather surprising conclusion and will be investigated further in the following chapters.

It is even more noteworthy that the modal split between car and rail—where the proportion travelling by rail increases as the trip length increases, as we showed in Table 4.11—appears not to depend on the relative journey speeds by car and rail, since these appear to be similar at all origin distances. As we shall see later, it has been a basic tenet of transportation modelling that the modal split is higher for a given mode the lower its relative costs (in time, money and effort). We shall see how this can be reconciled with our findings.

Over these two decades there has been a substantial change in the road network in certain parts of London, with a number of major radials being upgraded to dual-carriageway and even motorway

Table 4.14 Mean journey speed by distance from centre of origin, 1962
Independent regression equations

Sector 1	$v_c = 6.85\ (0.72) + 0.59\ (0.05)\ d$
	$v_r = 5.69\ (1.11) + 0.64\ (0.07)\ d$
Sector 2	$v_c = 6.79\ (0.94) + 0.78\ (0.08)\ d$
	$v_r = 4.97\ (0.77) + 0.73\ (0.06)\ d$
Sector 3	$v_c = 8.62\ (0.59) + 0.63\ (0.04)\ d$
	$v_r = 6.25\ (0.74) + 0.66\ (0.05)\ d$
Sector 4	$v_c = 7.15\ (1.14) + 0.71\ (0.08)\ d$
	$v_r = 5.46\ (0.69) + 0.73\ (0.05)\ d$
Sector 5	$v_c = 8.57\ (0.84) + 0.58\ (0.06)\ d$
	$v_r = 7.17\ (0.74) + 0.58\ (0.05)\ d$
Sector 6	$v_c = 6.35\ (0.53) + 0.68\ (0.04)\ d$
	$v_r = 5.58\ (0.72) + 0.60\ (0.04)\ d$
Total	$v_c = 7.74\ (0.46) + 0.63\ (0.03)\ d$
	$v_r = 6.18\ (0.44) + 0.63\ (0.03)\ d$
where	v_c = mean direct journey speed of cars
	v_r = mean direct journey speeds of rail
	d = mean distance of origin within a ring from centre (taken to be the midpoint of the ring)

The standard errors are in brackets

levels, starting with the opening of the M1 motorway at the beginning of the period.

We thought, therefore, it would be interesting to see if any effect of this increase in road capacity could be picked up in an analysis of journey speeds, though we were well aware that we would be stretching the statistical viability perhaps beyond its limits, owing to the small sample sizes in each category. Typically, the car categories would have, for example, only about 20 journeys in the sample.

We decided to perform an analysis of six 60° pie-shaped sectors, starting with a direction due east and running anti-clockwise as is the normal mathematical convention (rather than the geographical, which starts at north and runs clockwise). This is convenient in

Table 4.15 Mean journey speed by distance from centre of origin. 1971 Independent regression equations

Sector 1	$v_c = 7.20\ (1.33) + 0.53\ (0.09)\ d$
	$v_r = 5.69\ (0.67) + 0.61\ (0.04)\ d$
Sector 2	$v_c = 9.41\ (0.63) + 0.49\ (0.05)\ d$
	$v_r = 5.57\ (0.51) + 0.67\ (0.04)\ d$
Sector 3	$v_c = 9.16\ (1.23) + 0.59\ (0.09)\ d$
	$v_r = 5.79\ (0.57) + 0.71\ (0.04)\ d$
Sector 4	$v_c = 6.71\ (1.38) + 0.74\ (0.10)\ d$
	$v_r = 5.27\ (0.78) + 0.70\ (0.05)\ d$
Sector 5	$v_c = 9.76\ (0.95) + 0.47\ (0.07)\ d$
	$v_r = 6.86\ (0.46) + 0.59\ (0.03)\ d$
Sector 6	$v_c = 8.36\ (0.95) + 0.52\ (0.07)\ d$
	$v_r = 4.97\ (0.75) + 0.64\ (0.05)\ d$
Total	$v_c = 8.64\ (0.54) + 0.55\ (0.04)\ d$
	$v_r = 5.82\ (0.37) + 0.64\ (0.03)\ d$
where	v_c = mean direct journey speed of cars
	v_r = mean direct journey speeds of rail
	d = mean distance of origin within a ring from centre (taken to be the midpoint of the ring)

The standard errors are in brackets

London as the Thames estuary is due east of the centre. Having divided the data set into six, the standard errors will all increase by a factor of roughly 2.5.

The equations relating mean direct journey speed by car and rail to the distance of the origin from the centre are given in Tables 4.14 to 4.16 for each of the six sectors and for each survey. If one examines Blase's values, ie the cleaned values, for a particular sector (the South-West sector, sector 4, Figure 4.5), one can see that there is considerable variation in the values for a particular origin distance (or annulus) about the regression lines, as one would expect from the small samples sizes now being dealt with. This figure was also first published by Bly *et al* (1987).

Table 4.16 Mean journey speed by distance from centre of origin, 1981
Independent regression equations

Sector 1	$v_c = 5.43\ (0.96) + 0.66\ (0.07)\ d$
	$v_r = 5.74\ (0.56) + 0.52\ (0.04)\ d$
Sector 2	$v_c = 7.27\ (0.57) + 0.56\ (0.05)\ d$
	$v_r = 3.93\ (0.38) + 0.70\ (0.03)\ d$
Sector 3	$v_c = 7.65\ (0.86) + 0.56\ (0.06)\ d$
	$v_r = 3.69\ (0.58) + 0.73\ (0.04)\ d$
Sector 4	$v_c = 4.34\ (1.26) + 0.85\ (0.10)\ d$
	$v_r = 3.71\ (0.64) + 0.77\ (0.05)\ d$
Sector 5	$v_c = 7.87\ (0.71) + 0.49\ (0.05)\ d$
	$v_r = 4.25\ (0.60) + 0.67\ (0.04)\ d$
Sector 6	$v_c = 6.85\ (0.90) + 0.55\ (0.07)\ d$
	$v_r = 3.64\ (0.37) + 0.69\ (0.24)\ d$
Total	$v_c = 6.66\ (0.43) + 0.60\ (0.03)\ d$
	$v_r = 4.44\ (0.32) + 0.65\ (0.02)\ d$
where	v_c = mean direct journey speed of cars
	v_r = mean direct journey speeds of rail
	d = mean distance of origin within a ring from centre (taken to be the midpoint of the ring)

The standard errors are in brackets

Bly *et al* (1987, p.9) go on to note that:

> 'In sectors whose rail travel was faster than average car travel was also faster, suggesting an underlying connection between the speeds on the two modes, but in general the variability of average speeds in different annuli and different sectors does not suggest that there is any systematic equality'.

There is, however, one method whereby the systematic variation can be examined more closely. It is evident from the equations for

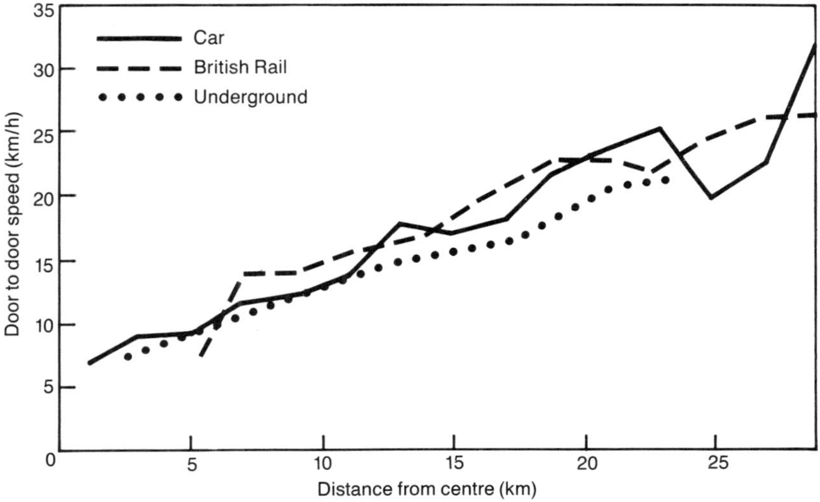

Figure 4.5 Direct journey speed by distance of origin from centre, morning peak-hour (7–10 am), central destinations within 6-km of centre, GLTS 1981, sector 4

car and rail presented in Tables 4.14–4.16 that, owing to the estimation process, a high intercept is likely to be associated with a low slope coefficient and vice versa.

If we therefore estimate the average direct speed for an origin distance of 15 km, which is about the midpoint of the distance range examined, then we find the data shown in Table 4.17.

It can be seen from this table that in almost all cases the 1962 and 1971 mean speeds are very similar or fall slightly over the decade, but fall more steeply to 1981. Sectors 2, 3 and 4 are above average; 1, 5 and 6 are below average in general.

This analysis does, therefore, suggest to me that there is a systematic equality, or rather similarity, given that these values are from the uncleaned data.

It will be noted that it is precisely in those directions from which the LT tube network is better that the higher speeds of travel are found, and in those directions from which the BR network is the main provider of rail service and the tube network is poor that the lower speeds of travel are found.

Table 4.17 Mean direct journey speed by sector by car and rail for journeys beginning 15 km from centre, morning peak-hour (7–10 am), central destinations within 6-km of centre

sector	mode	1962 (km/h)	1971 (km/h)	1981 (km/h)
1	car	15.70	15.15	15.33
	rail	15.29	14.84	13.54
2	car	18.49	16.76	15.67
	rail	15.92	15.62	14.43
3	car	18.07	18.01	16.05
	rail	16.15	16.44	14.64
4	car	17.80	17.81	17.09
	rail	16.41	15.77	15.26
5	car	17.27	16.81	15.22
	rail	15.87	15.71	14.30
6	car	16.55	16.16	15.10
	rail	14.58	14.57	13.99
total	car	17.19	16.89	15.66
	rail	15.63	15.42	14.19

In general, these data are not adequate, as we feared, to enable us to relate road or rail changes in a sector to the change or otherwise in the equations relating speed and distance.

It is worth noting at this point that the 1962 LTS (LTS Volume 1, 1964) gave a diagram of the travel time contours by car in the off-peak from Charing Cross to all parts of London, the first time that such geographical representation had been done. This is reproduced in Figure 4.6. The contours are given at 10-minute intervals. Paragraph 2.66 states that these contours are based on 'field observations'!

It can be seen that the speed of travel in the centre is as low as six miles/h in some directions. The speed gradually increases as the distance from the centre increases, and is generally circularly symmetric, though slightly slower to the north-east and faster to the south-west, as we have found in the journey speed data. This gives a

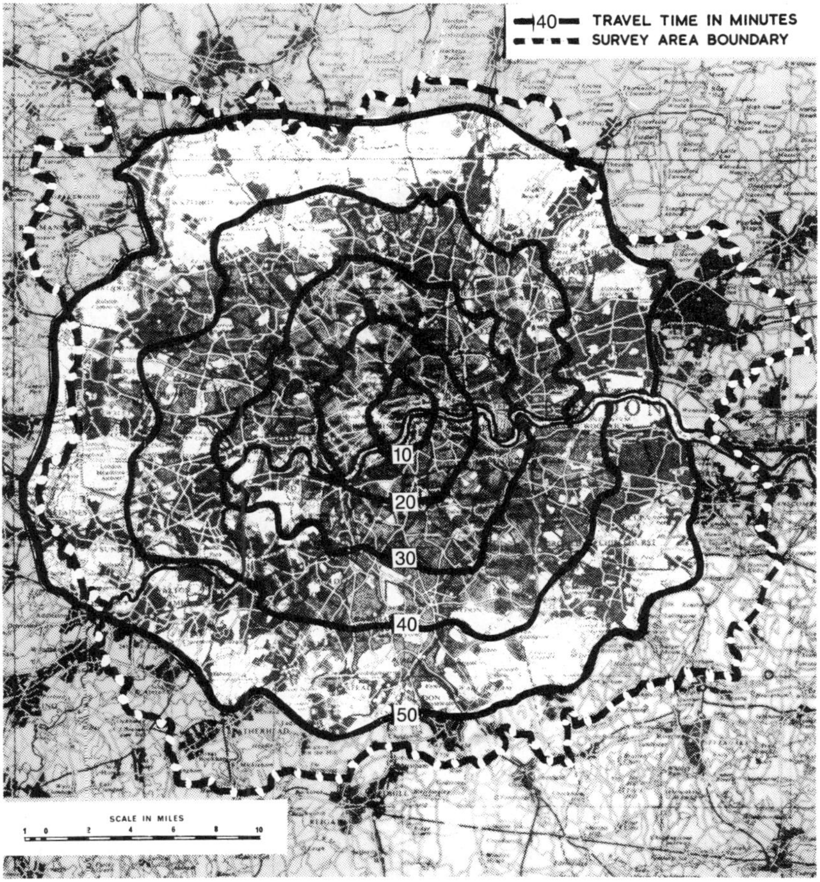

Figure 4.6 Off-peak travel time contours by car LTS 1962

slightly egg-shaped contour, with the point top to the south-west and the blunt bottom to the north-east.

Paragraph 2.66 states:

> 'To reach the External Cordon from the centre of London generally requires 45 to 55 minutes even in off-peak traffic periods and 55 to 65 minutes in peak periods'.

As can be seen from the journey speeds in Table 4.17, this was somewhat optimistic, since in 1962 a 15 km (nine mile) journey had a journey speed of 17.19 km/h by car, thus taking 52.35 minutes in the peak. Part of the reason is undoubtedly the difference between kerb-to-kerb speed and total journey speed, ie the access times.

Paragraph 2.67 of the LTS Volume 1 is worth quoting in full in that it shows the then perception of travel times.

> 'One of the reasons for the increasing use of peak period rail services to Central London is the fact that the railways can offer significant travel time advantages over road transport for many of the longer journeys of this type. Journeys to Central London from the outer fringe of the Survey Area can generally be accomplished in 10 to 15 minutes less time by railway than by car, including waiting time and travel time to and from the stations. In general, travel time advantages of this magnitude and convenience factors appear to be more significant in mode selection than relative costs.'

Thus at this point in time rail was seen to be faster than car for many of the longer journeys to the centre, even when access was included. This was not correct, because access was underestimated. Whilst rail running speeds are considerably faster than car running speeds (or kerb-to-kerb speeds), the access times to stations in outer London are considerably more than in inner London.

We can show this by an analysis of the 1971 GLTS data for central area destinations. We reduce the peak from 7–10 am to 8–10 am arrivals, and we limit the analysis to those journeys for which the access components are available. Not all journeys give access components. On the other hand, we expand the analysis to differentiate between travellers with a car available and those without, to see if there is any difference in the way that the rail system is used. In particular, we wished to find out whether travellers without a car were prepared to travel further to a rail station than those with.

We also wished to find the size of the catchment areas surrounding stations, and to compare it with data determined more directly by other surveys, which we will present afterwards. It must be noted that the access distance in the GLTS is only available from or to the zonal centroid and the rail station, not the precise address. Whilst this will not matter much in the central area, it may cause problems for origin access as zonal sizes increase as the distance of the origin from the centre increases.

In the central area, the distributor part of the journey will include the LT component of a BR journey, but will not include any transfers within a LT journey, since these are not recorded. Thus some of the central access could itself be by rail, if the trunk-haul part of the journey is by BR. This is a peculiarity of the way the data is recorded.

The results are presented in Tables 4.18 for origin access and 4.19 for destination access. The complementary analysis of the trunk-haul component is not given here.

We have used simple least squares regression analysis as before on the direct lengths, times and direct speeds of the access components at each end of the journey for no-car available and car available travellers with respect to the distance of the origin from the centre. These are given in Table 4.20, together with the regressions on the total time. The rail running time equation is then simply obtained by subtraction. One would expect the rail running time equations to be identical for car-available and no-car-available travellers.

The regression equations for total time are almost identical and well within standard errors. They imply a marginal total direct speed of 33.3 km/hour at all distances of the origin from the centre. Those for running time have very similar slope coefficients, implying a marginal running direct speed of 60 km/hour at all distances. These results are slightly misleading in that the regressions are taken against the distance of the origin from the centre, rather than the actual distance travelled.

The differences between these two imply that as the distance travelled increases, then so too do the access times. The equations for the total direct speeds given earlier in Tables 4.12 and 4.13 can be shown to be similar to those just given with appropriate transformations of variables, but they are not strictly comparable, since the total distance here is the sum of the direct distances of each of the components, rather than the true direct distance between origin and destination.

The direct running time equation implies that average running speed increases as the distance of the origin from the centre

Table 4.18 Analysis of direct distance, travel time and direct speed of the access components of rail journeys from the origin in the morning peak-hour 1971 GLTS sample of journeys by distance of origin from centre, central area destinations between 8 – 10 am

origin ring (km)	*number*	*direct distance (km)*	*access time (mins)*	*direct speed (km/h)*	*running time (mins)*	*total time (mins)*
i) no car available						
0 – 2	95	0.00	5.00	0.00	8.00	15.00
2 – 4	340	0.17	9.01	1.12	9.68	27.68
4 – 6	3357	0.63	14.10	2.66	9.91	37.11
6 – 8	8586	1.07	14.98	4.29	13.41	40.28
8 – 10	10189	1.14	14.38	4.74	15.76	46.01
10 – 12	13150	1.75	18.70	5.60	18.12	53.26
12 – 14	10112	1.55	16.92	5.51	20.81	51.84
14 – 16	9962	1.64	17.62	5.58	21.38	55.99
16 – 18	9491	1.92	20.50	5.62	22.82	60.29
18 – 20	6440	2.59	20.84	7.47	25.66	64.99
20+	13757	2.37	22.29	6.37	27.44	67.63
ii) car available						
0 – 2	0	0.00	0.00	0.00	0.00	0.00
2 – 4	257	0.51	7.83	3.92	6.20	20.70
4 – 6	2851	0.68	11.04	3.69	10.77	34.16
6 – 8	7483	0.96	14.52	3.96	13.37	40.39
8 – 10	11314	1.03	13.90	4.43	16.83	44.68
10 – 12	15416	1.32	16.35	4.83	18.02	50.01
12 – 14	18715	1.45	14.16	6.15	22.57	52.04
14 – 16	20394	1.93	17.24	6.72	23.36	56.29
16 – 18	30398	2.07	17.14	7.26	25.15	58.57
18 – 20	26268	2.39	18.66	7.67	27.91	63.30
20+	51990	2.75	20.21	8.17	28.38	66.57

Table 4.19 Analysis of direct distance, travel time and direct speed of the access components of rail journeys to the destination in the morning peak-hour 1971 GLTS sample of journeys by distance of origin from centre, central area destinations between 8 – 10 am

origin ring (km)	*number*	*direct distance (km)*	*access time (mins)*	*direct speed (km/h)*	*running time (mins)*	*total time (mins)*
i) no car available						
0 – 2	95	0.00	2.00	0.00	8.00	15.00
2 – 4	340	0.88	8.99	5.86	9.68	27.68
4 – 6	3357	1.20	13.11	5.51	9.91	37.11
6 – 8	8586	1.19	11.76	6.07	13.41	40.28
8 – 10	10189	1.70	15.87	6.43	15.76	46.01
10 – 12	13150	1.62	16.44	5.92	18.12	53.26
12 – 14	10112	1.49	14.19	6.30	20.81	51.84
14 – 16	9962	2.12	16.99	7.49	21.38	55.99
16 – 18	9491	2.04	16.96	7.23	22.82	60.29
18 – 20	6440	2.46	18.43	8.01	25.66	64.99
20+	13757	3.17	17.82	10.69	27.44	67.63
ii) car available						
0 – 2	0	0.00	0.00	0.00	0.00	0.00
2 – 4	257	0.73	6.66	6.54	6.20	20.70
4 – 6	2851	1.54	12.34	7.49	10.77	34.18
6 – 8	7483	1.20	12.50	5.75	13.37	40.39
8 – 10	11314	1.28	13.95	5.52	16.83	44.68
10 – 12	15416	1.81	15.69	6.92	18.02	50.01
12 – 14	18715	1.73	15.34	6.77	22.57	52.04
14 – 16	20394	1.77	15.69	6.75	23.36	56.29
16 – 18	30298	1.80	16.28	6.63	25.15	58.57
18 – 20	26268	2.11	16.73	7.56	27.91	63.30
20+	51990	2.52	17.97	8.43	28.38	66.57

Table 4.20 Regression equations of mean direct distance, time and direct speed of access components of rail journeys by distance of the origin from the centre, for car available and no-car available travellers

i) origin access	l_c	=	$-0.22\ (0.17) + 0.14\ (0.01)\ d$
	l_0	=	$0.86\ (0.27) + 0.10\ (0.02)\ d$
	t_c	=	$9.9\ (1.3\ ) + 0.47\ (0.08)\ d$
	t_0	=	$11.0\ (2.0\ ) + 0.53\ (0.13)\ d$
	s_c	=	$1.89\ (0.40) + 0.30\ (0.02)\ d$
	s_0	=	$3.53\ (0.63) + 0.14\ (0.04)\ d$
ii) destination access	l_c	=	$0.56\ (0.26) + 0.09\ (0.02)\ d$
	l_0	=	$0.27\ (0.41) + 0.12\ (0.03)\ d$
	t_c	=	$11.2\ (1.0\ ) + 0.31\ (0.06)\ d$
	t_0	=	$11.6\ (1.6\ ) + 0.31\ (0.10)\ d$
	s_c	=	$4.17\ (0.79) + 0.19\ (0.05)\ d$
	s_0	=	$3.05\ (1.24) + 0.31\ (0.08)\ d$
total time	T_c	=	$29.0\ (1.52) + 1.79\ (0.09)\ d$
	T_0	=	$29.8\ (2.37) + 1.81\ (0.15)\ d$
running time	r_c	=	$7.9 + 1.01\ d$
	r_0	=	$7.2 + 0.97\ d$
where	l	=	average direct access distance (km)
	t	=	average access time (mins)
	s	=	average direct speed (km/h)
	T	=	average total journey time (mins)
	r	=	average running time (mins)
	d	=	distance of origin from centre (km)
	subscript c		refers to car available
	subscript 0		refers to no-car available

The standard errors are in brackets

increases, being about 35 km/h at 10 km from the centre and 44 km/h at 20 km.

In fact, if we performed a regression on the running time against the direct distance of the running component we would find that average running speeds increase from about 20 km/h at the centre to 36 km/h at 20 km from the centre. These values are slightly lower because the direct distances travelled on the running component are lower than the distance to the centre. These values of average direct running speed against distance of the origin from the centre are shown in Figure 4.7 for car-available and no-car-available travellers.

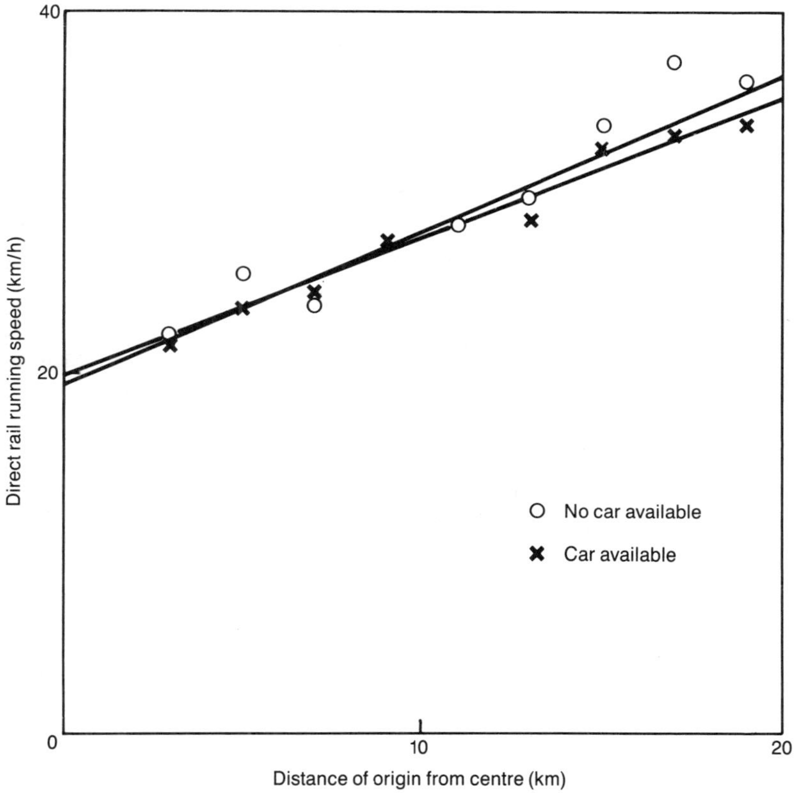

Figure 4.7 Average direct running speeds for the running component of rail journeys to the central 6-km circle for car-available and no-car-available travellers by the distance of the origin from the centre, morning peak 8–10 am, GLTS 1971

The regression equations for average direct speeds for the running components for car-available and no-car-available travellers are as follows:

$$v_c = 19.9 + 0.77\,d$$
$$v_0 = 19.5 + 0.85\,d$$

where v_c = direct speed of running component for car-available travellers

v_0 = direct speed of running component of no-car-available travellers

d = distance of origin from centre

This is what one would expect, as the distances between stations increase as the distance from the centre increases, giving faster local rail speeds and faster average rail speeds to the centre. The local rail speed increases from about 20 km/h in the centre to about 52 km/h (20 + 2*(36–20)) at 20 km from the centre.

The values of 19.9 or 19.5 km/h (12.3 and 12.1 miles/h respectively) are very close to the value of 12.3 miles/h already quoted after the analysis of Table 3.11 for the rail direct running speed in the centre.

The value of 36 km/h (22.3 miles/h) had, it will be remembered, already been achieved for a number of services from comparable distances from the centre in 1904 as shown in Table 3.14., though these quoted values are measured along the track rather than in direct lines.

With these direct running speeds, which are much faster than the equivalent road speeds, it is perhaps not surprising that the LTS should have stated, as quoted above (LTS, 1964, paragraph 2.67), that:

> 'the railways can offer significant travel time advantages over road transport for many of the longer journeys of this type'.

What they did not know was that access times increase as the distance of the origin from the centre increases, and are greater than assumed in the LTS.

At the destination end (i.e. in the central circle) all the pairs of equations for car-available and no-car-available travellers are not significantly different. Interestingly, the direct speeds are a little higher at the destination end than at the origin end; this may be due to the inclusion of LRT tube as access for BR rail trunk-haul trips, or it might simply be due to better bus provision in the central area. Walking would give a direct speed of about 2.5 km/h. The direct speeds of 5–10 km/h are somewhat higher than could be achieved just by walking, and are in fact comparable to the speeds achieved on actual complete journeys entirely within the central area, though their direct length and time is somewhat shorter.

It is worth noting that central area access time is estimated at just over 11 minutes for an origin at the centre, rising to 17 minutes for peripheral origins at 20 km. This value of 11 minutes for an origin at the centre is, it will be remembered, somewhat higher than the values given in Tables 3.4 and 3.9 for the walk-to-stop components of journeys by bus (about five minutes) and rail (about seven to eight minutes) entirely within the central 10-km square.

This current analysis of actual journeys, however, is by main mode, so that walk, bus and both LT and BR rail could be subsidiary modes of a BR or LT rail main mode journey; this is quite different to the potential random journeys described earlier, where only walk is measured separately.

At the origin end, it can be seen that the equations giving the direct length are different for car-available and no-car-available travellers, though these equations are subject to error from the use of the zonal centroid rather than a precise address as an origin. There is no significant difference in the time equations. The direct speed equations are also different, car-available travellers having lower direct speeds nearer the centre, but rising more rapidly with distance of the origin from the centre than no-car-available travellers.

Essentially, although the linear equations distort the results, this suggests that car-available rail travellers travel more by faster access modes as the distance of the origin from the centre increases, and in particular by bus and car, rather than by walk.

The difference is not great, however. At 20 km from the centre, the equations suggest direct distances of 2.6 km and 2.8 km by car-available and no-car-available respectively, (i.e. the former are shorter), with times of 19.3 minutes and 21.6 minutes and direct speeds of 7.9 km/h and 6.3 km/h.

I suspect that in the LTS, access times of 10 minutes were chosen to represent access at the origin. Whilst this is approximately correct for central area origins, it underestimates access times at peripheral (i.e. 20 km) origins by 10 minutes, leading to the mistaken conclusion already quoted (LTS, 1964, paragraph 2.67), namely:

> 'Journeys to Central London from the outer fringe of the Survey Area can generally be accomplished in 10 to 15 minutes less time by railway than by car, including waiting time and travel time to and from the stations'.

This conclusion should have been that journey times by car and rail to central London from the periphery, or outer fringe, are similar on average.

I have shown that car-available travellers who choose to travel by rail live slightly closer to rail stations than no-car-available travellers, but the differences are not large in the periphery. This fact will be important when we come to look at the theory.

More importantly, the average direct access distance travelled is comparable to the spacing between rail stations, implying many travellers travel to stations further away than their nearest station, presumably in order to minimize total travel time by reducing destination access time.

We can show this more directly by an analysis of the locations of addresses of people who use particular rail stations. As part of the 1981 GLTS, British Rail and London Transport carried out a sample survey of passengers entering their stations. They have both kindly provided me with maps showing the locations of the origins of journeys by their passengers on two adjacent lines in North London, the Great Northern of British Rail and the Piccadilly of London Transport.

These maps were prepared to my specification under the direction of Mr Lee, British Rail's Network South-East Market Analysis Manager and Dr Fairhurst, London Transport's Deputy Group Planning Manager, so that I never saw the original confidential information of passengers' addresses. I specified that addresses in a given road should be spread out equally along the given road.

The first figure, Figure 4.8, shows the locations of these two lines in relation to neighbouring lines in North London, and their penetration of the centre of London. The Great Northern splits into a Y-shape with the eastern arm of the Y providing only an inner suburban service. The western arm has a four-track line, with the outer two tracks providing an inner suburban service and the inner two tracks, the main line, providing non-stop outer suburban and inter-city services which are of no concern to us here. The Piccadilly line splits the Y-shape down the middle.

The two lines on either side of our study area are the High Barnet branch of the Northern line of London Transport to the west, and the Enfield line of British Rail to the east.

The locations of a sample of the travellers who travel to two particular adjacent stations in the study area are given in Figure 4.9.

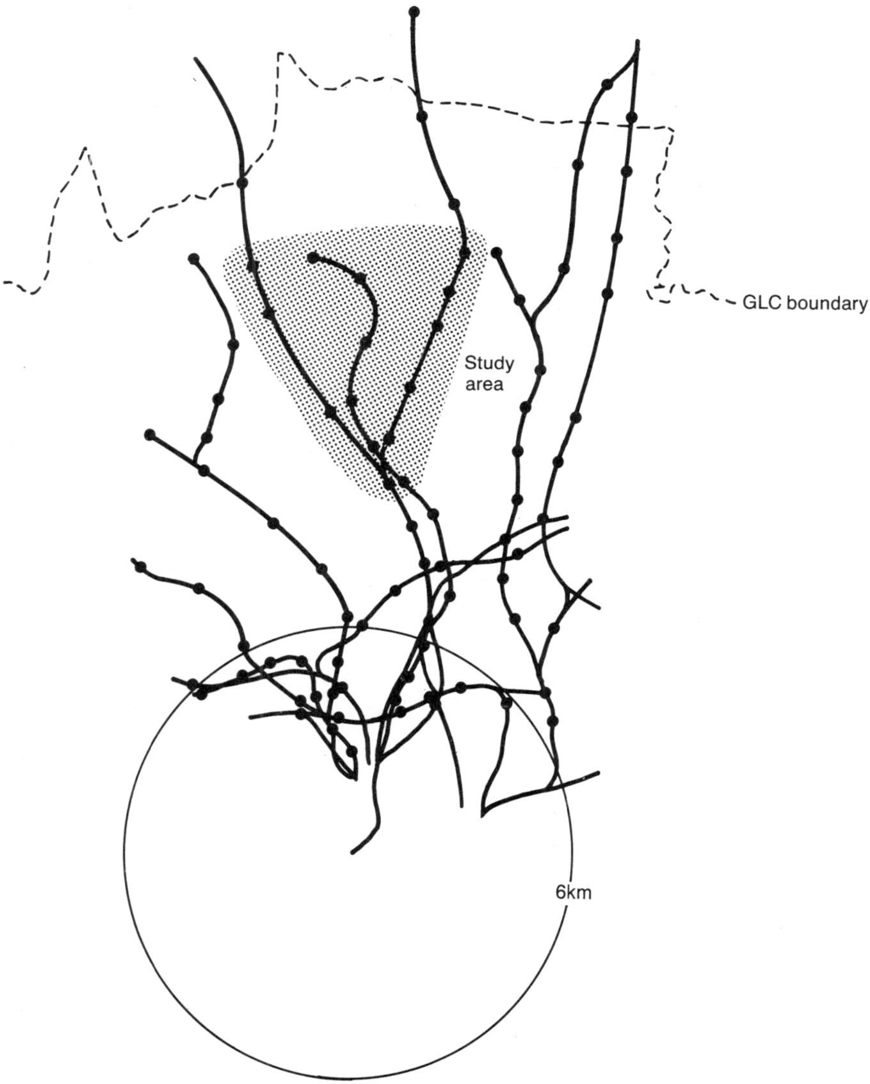

Figure 4.8 The location of rail lines in North London

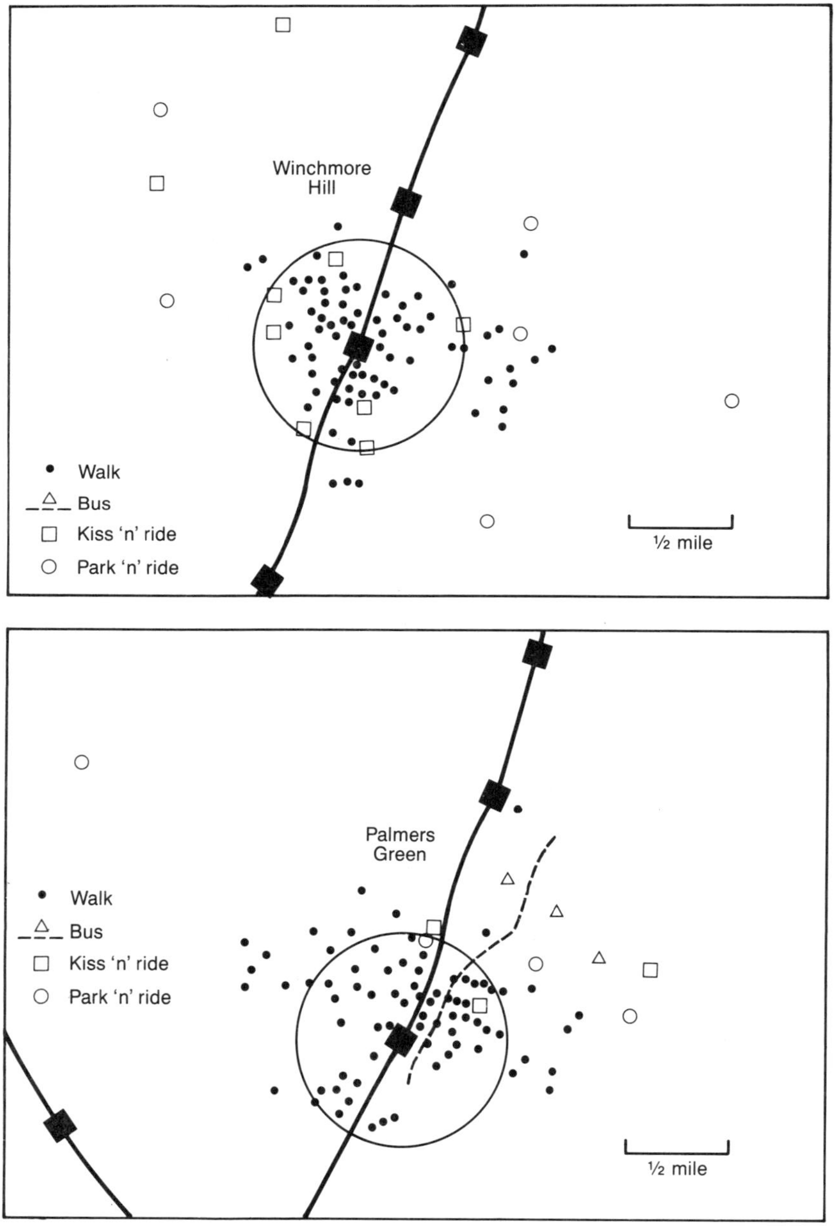

Figure 4.9 Location and method of access of travellers to two adjacent stations on the same line in North London

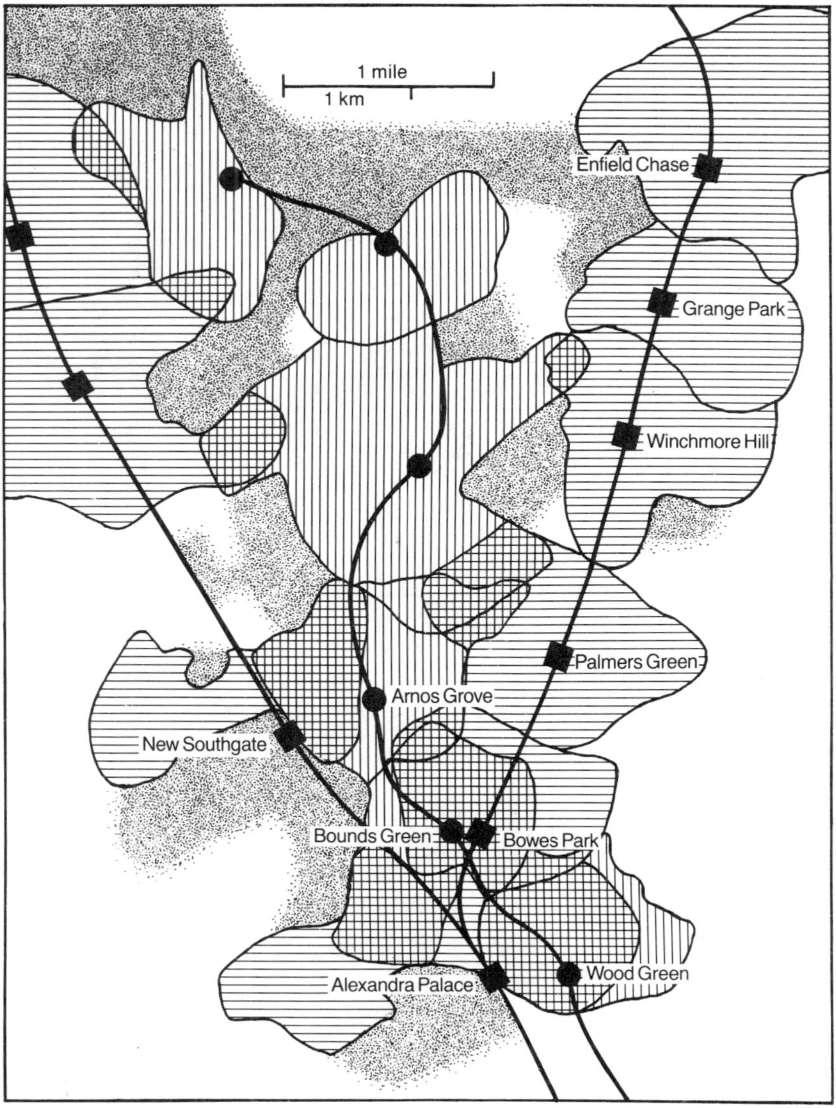

Figure 4.10 Catchment areas of railway stations in the North London outer suburbs

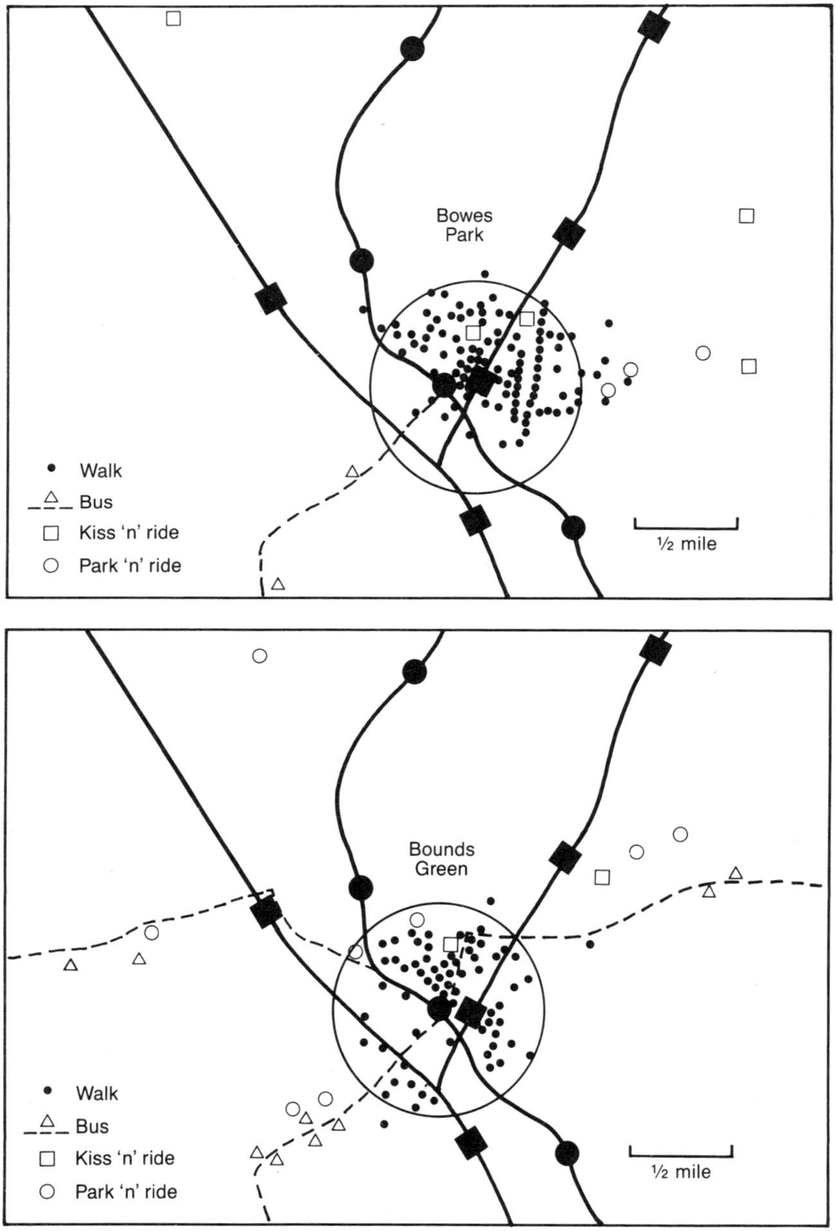

Figure 4.11 Location and method of access of access of travellers to two neighbouring stations on different lines

Whilst the vast majority walk to the stations, a sprinkling of travellers by bus, kiss-and-ride and park-and ride can also be seen. Interestingly, there appear to be walkers who walk to a station further down the line rather than to the adjacent station, which probably implies some element of deliberate choice to walk a certain distance for health reasons.

The catchment areas of the stations in our study area are shown in Figure 4.10. These lines are drawn around all passengers who walk to the stations; some outlying passengers who travelled by bus or car are not included.

The non-residential areas are given as open space or other (including industrial). It can be seen that the catchment areas completely cover the area, except for the non-residential areas. There is, however, some overlap, particularly where the lines become closer together towards the south. We show this in more detail by examining the catchment areas of Bowes Park and Bounds Green, two stations within a few yards of each other on the Great Northern and Piccadilly respectively: Figure 4.11. These two stations are about 10 km from the centre.

It can be seen that these catchment areas overlap considerably. Obviously, travellers are in general choosing the appropriate station to minimize their total journey time, including the access at their destination. The Piccadilly line station at Bounds Green also has an overlap to the south with the Great Northern station at Wood Green (now renamed Alexandra Palace). The circles are drawn with a half-mile radius.

It can also be seen that the vast majority of travellers walk to the station, with only a few of the relatively longer trips being made by bus or car, either kiss-and-ride or park-and-ride. With the limited car parking space available at most of these stations, there is very little park-and-ride. Further out, and at particular stations, there is considerably more parking available.

The catchment areas presented here seldom stretch much more than one km from the station; for a distance from the centre of 10 km, we would expect from the analysis above an access time of about 16 minutes, which is about 2/3 km direct distance walking at a speed of 2.5 km/h. The outliers of travellers by bus and car do, however, bring up the average distance and speed of travel.

This detailed picture of origins does then confirm the picture from the GLTS zonal analysis. If we add up access at both ends of the rail journey, access for journeys can be seen to be some 2/3 of total journey time for central origins, and only falls to some 60 percent of the total for origins on the periphery.

We return now to complete our discussion of the variation in rail journey times begun in the previous chapter by examining some journeys from outer London to the Transport Studies Group. Three of my colleagues recorded their journeys during 1987, of which one continued recording to the end of 1989.

The journeys are as follows in order of direct distance as before, and are shown in Figure 4.12.

1) Kay. *Brentwood–Kings Cross.* Kay interchanges at Liverpool Street onto the Metropolitan and Circle Lines. He continued to record his journeys until the end of 1989, so it is possible to split his journeys by year. This is done at the change of timetable in May;

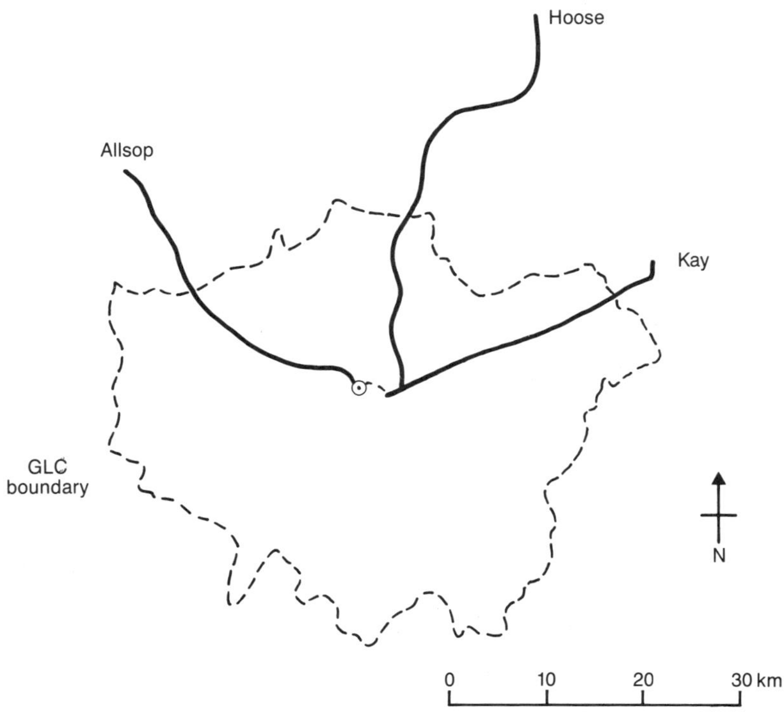

Figure 4.12 Rail journeys from outer London (1987)

normally, this would not be expected to have any effect, but in May 1988 a substantial down-grading of the inner suburban services to Brentwood was made for the benefit of the longer outer suburban services to Essex and East Anglia.

Kay recorded a median travel time for 77 journeys up to May 1987 of 56.5 minutes; this decreased slightly for 174 journeys to 53.8 minutes in the year 1987–88; after the change of service, this increased for 166 journeys to 63.0 minutes in the following year, though they have since improved slightly for a further 110 journeys to 59.8 minutes in the final period of 1989. Kay changed his place of work in June 1989; this added two minutes to his rail journey from Kings Cross to Euston Square but did not affect his access time. We have simply subtracted the two minutes from the later journey times to give a consistent series.

It should also be noted that Liverpool Street station was undergoing modernization during this period, and substantial problems occurred during resignalling. 1989 was also notable for a series of rail strikes, both on BR and LUL, which have been ignored in this analysis.

2) Allsop. *Hemel Hempstead–Euston.* Allsop recorded 82 journeys in the peak hour during 1987. Although he has a direct route, the trains vary in the number of intermediate stops from 0 to 4, so the coefficient of variation is higher than it would be if the number of stops were always the same.

3) Hoose. *Bishop's Stortford–Kings Cross.* Hoose recorded 75 journeys up to the change of timetable in May 1987, with a median time of 88.3 minutes, and a further 119 journeys on the new timetable up to March 1988 with a median time of 76.4 minutes. He interchanges at Liverpool Street, catches the same train every day and now has one stop between Bishop's Stortford and Liverpool Street. It can be seen therefore that Hoose has benefitted from the change of emphasis of British Rail from inner to outer suburban services whereas Kay has correspondingly suffered.

The cumulative logarithmic normal distributions of travel time are shown in Figure 4.13 as before, and the estimated medians, direct speeds and proportion of travel time in access in Table 4.21.

It can be seen that Allsop, though he spends a much higher proportion of his journey in access, has a faster direct journey speed, mainly because he does not have to interchange to a slow distribution

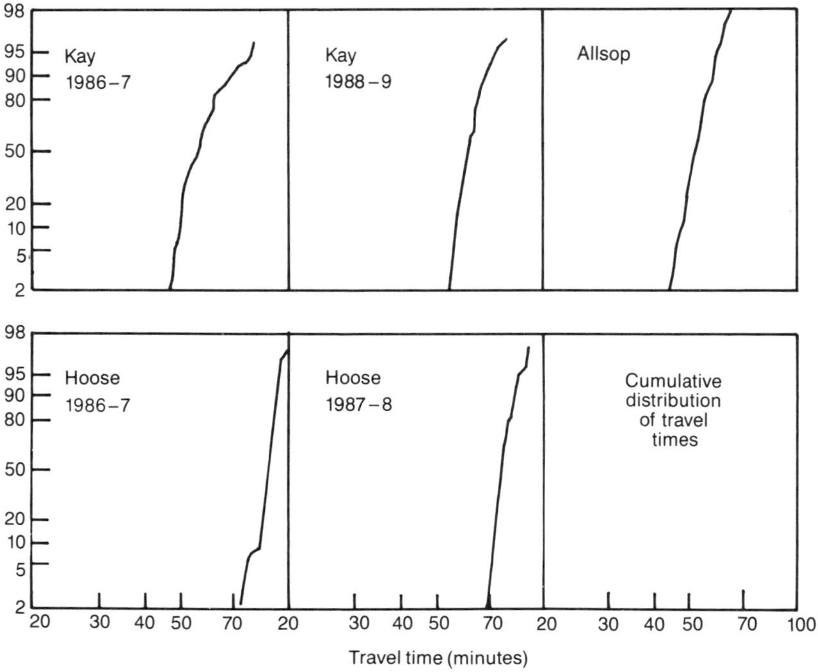

Figure 4.13 Cumulative distributions of travel time for outer London rail journeys

Table 4.21 Rail journeys in outer London (1987–89)

	direct dist: miles	*median time mins.*	*direct speed mi/h*	*coeff of vari:*	*access time mins.*	*percent in access*
Kay						
(1987)	31	56.5	32.9	0.15	9	16
(1987–8)		53.8	34.6	0.08		17
(1988–9)		63.0	29.5	0.08		14
(1989)		59.8	31.2	0.09		15
Allsop	34	53.4	38.4	0.10	18	34
Hoose						
(1987)	43	88.3	29.2	0.04	14	16
(1987–8)		76.4	33.8	0.04		18

mode, the LUL rail system. If we approximate Allsop's running time by subtracting access, he has a direct median running speed of 57.6 miles/h. Kay's earliest results give a median direct running speed of 39.2 miles/h, and Hoose's earliest give 34.7 miles/h, though their latest results reverse this order, being 36.6 and 41.3 miles/h.

If we use the equation in Table 4.20 for direct running times for car owners, then the average direct running speeds for distances of 34 miles, 31 miles and 43 miles should be 32.3, 31.9 and 33.1 miles/h. This is, not surprisingly, wrong because we are out of the valid range; this equation gives direct running speeds which asymptotically approach 60 km/h (37 miles/h) as distances become larger, and thus underestimate the speeds at the longer distances.

Finally in this chapter, we present an analysis (Beardwood, 1989) of the access times for car travel in the centre of London taken from the 1981 GLTS survey. Central London is here defined as the conurbation centre, not our six-km radius circle, so is more tightly drawn round the centre. Because there are very few car journeys in the sample to the centre, we group the origins by central, inner and outer London, and external to the survey area (these latter are trips recorded by car drivers with a home inside the survey area). These are the raw data, and are all-day values.

Most car journeys do not record access times, and only a minority record access times at both origin and destination. We therefore separate off journeys with access recorded at each end from those with no access recorded, to see if there is any difference in the total journey time. Table 4.22 gives the results.

These values shown in Table 4.22 suggest that pre- and post-access times taken together, where recorded, are about 10 minutes for journeys entirely in central London, and rise slightly with distance to about 14 minutes for journeys to the centre from outer London. The vast majority of trips, however, record no access times; for journeys within central London and from inner London, the journey times are significantly lower than those where access is recorded, although not always for those from outer London.

Obviously more careful work is necessary to confirm the true value of car access times in central London, but it does at least seem possible that a value of access time of 10 minutes in a 20-minute journey is not impossible. Values of this order would enable the direct speeds door-to-door of central area car journeys measured in

Table 4.22 Access times for car trips to central London 1981 GLTS

			origins		
	central	*inner*	*outer*	*external*	*total*
pre-access=0, post-access=0					
running time	20.6	29.7	50.7	110.5	35.0
journey time	20.6	29.7	50.7	110.5	35.0
sample size	599	1081	798	24	2502
pre-access=0, post-access>0					
running time	15.6	32.1	52.2	158.3	39.8
post-access time	12.4	9.5	12.9	28.3	11.5
journey time	28.0	41.6	65.1	186.6	51.3
sample size	75	253	277	3	608
pre-access>0, post-access=0					
pre-access time	9.2	4.7	14.8	-	7.8
running time	13.4	33.5	48.5	—	24.5
journey time	22.6	38.2	63.3	—	32.3
sample size	81	64	13	—	158
pre-access>0, post-access>0					
pre-access time	5.1	4.2	6.1	1.0	4.8
running time	16.8	26.1	38.3	28.0	24.0
post-access time	5.3	8.6	7.0	5.0	7.0
journey time	27.2	38.9	51.4	34.0	35.8
sample size	55	59	19	1	134

the GLTS surveys to be reconciled with the traffic speeds measured in Chapter Two.

We turn now to examine the history of transportation modelling, to see why the perceptions about the importance of public and private transport developed in the way that they have and why no such measurements as have been presented above about journey speeds were available before our own examination of them.

5 *Transportation Models*

Although transportation modelling in London and the UK begins with the London Traffic Survey of 1962, the collection of the kind of travel data required for the modelling process predates this study by more than a century.

Essentially in a transportation model we are attempting to relate the data about the network with the data about how individuals are using the network, both for transporting themselves and for moving goods. The aim is to replicate the vast amount of data with some relatively simple relationships. Once this is done, future travel demands can be assessed against possible future networks. The best investment and management strategies can then be chosen.

We have already examined the history of the travel speed data. We now turn to the history of the data about individual journeys.

Origin-destination surveys in fact go back a long way in London. In the evidence presented to the Royal Commission on Metropolis Railway Termini of 1846, we see a presentation by Horne of the origins and destinations of passengers by omnibus and cab at Euston (the London and Birmingham Railway terminus), gathered over a month, and the change in relative times and costs that would be involved if the terminus were moved to the proposed site at Farringdon Street in the City. This is noted in Prest and Turvey's survey of cost-benefit analysis (1965) and in Kellett (1969, 1979, p.40).

The Royal Commission on London Traffic in 1906 gave a set of times and costs that had been measured from various homes to given workplaces (Volume 8, Table 8) and, as already noted in the note following Table 3.15, had added ten minutes walk at each end of the rail journey to represent access for a door-to-door journey.

We also have a claim, in 1936, to have performed the first origin-destination survey by Bressey and Lutyens (1938, paragraphs 28ff.). This was a survey of the goods traffic from the London docks gates.

All these pre-date the Toledo origin-destination survey of 1942, which Barkley (1950), repeated in Vaughan (1987), claims as the first such. Barkley also gives the Chicago cordon count of traffic of 1916 as being the first such. In London, the House of Commons Select Committee on Metropolitan Communications of 1854–55 had data on traffic flows into the City presented to it by Pearson, Solicitor to the City Corporation. These data are given in Barker and Robbins (1963) and Kellett (1969, 1979 p. 45). And in the 1906 Royal Commission we find a count of the number of passengers coming into London each day, as noted in Mogridge (1988a) and Department of Transport (1989b).

The first household survey in London was the survey carried out in 1949 by Research Services Ltd for the London Transport Executive (LTE, 1950), then a newly regenerated body under the British Transport Commission (BTC) formed by the 1947 Act. It was the first household survey of travel in the UK and, as such, experimental.

It covered 3,000 households selected in an area a little bigger than the London conurbation definition of the Registrar-General. Interviews were carried out in the period 24 January-14 April 1949. Only journeys made by public transport were recorded. Whilst information was published on in-vehicle journey time (Table 12) and total journey time to work (Table 13), no analysis was made of journey speed.

The second household survey in 1954 (LTE, 1956) was carried out by the Social Survey on behalf of the Road Research Laboratory (RRL), the BTC and the LTE. A slightly larger area was sampled, doubling the number of nearby towns in the 1949 survey from three to six, but the sample size was increased from roundly 2,700 to 4,000 households in the conurbation area and from 300 to 1,500 in the adjacent towns.

Data was collected of all journeys in the previous seven days, rather than that of 'regular' and 'casual' journeys as in the first survey, and in particular covered all journeys, not just those made by public transport. Interviews were carried out between 26 January and 14 April 1954 in the conurbation, and up to 5 June elsewhere. The response rate was roundly 70 percent.

Although both the times of journeys and the grid reference of the origin and destination of each journey were obtained, to a half-mile grid, the published results only include the total journey time to work for those using public transport (Table 10), giving a value (39 minutes) very similar to that of the 1949 survey (42 minutes) (conurbation resident workers only).

Paragraph 46 of the report (p.21) does, however, make a comment about car journey speeds. Since the time spent in parking and on walking to the place of work were not recorded, it is difficult to know what weight to give the observation that:

> 'for the generality of journeys, the private car seems to have been quicker than public transport'.

The direct distances were later analysed in Dawson and Wardrop (1962) for journeys by road only. They took, quite reasonably, the centre points of the grid square in which the trip end lay as the definition of the trip end, and they defined trips which begin and end in a grid square as having zero mileage. This does, however, mean that the resulting average direct distances travelled are likely to be biased.

This is particularly so for bus trips where the average direct trip length quoted (Table 15: 1.7 miles) is comparable to the grid square dimension (0.5 miles). Car trip lengths are quoted as 3.4 miles on average with both trip ends in the Greater London area (as defined by a large grid square—not by the administrative boundary of the survey itself).

These values are not unfortunately tied to comparable values of journey times.

These latter two household surveys, being pre-computer, were all analysed from punched cards. I have not been able to track down what happened to those boxes of cards as yet.

The next household survey by the Social Survey was that of Gray's (1969) *Private Motoring in England and Wales* from 1960 to 1964. Whilst this is not a London survey as such, London formed a sub-set of the analysis. The results, however, not being available until 1969, did not affect the initial manner in which the London Traffic Survey of 1962 was analysed, and so will be discussed further in Chapters Seven and Nine, together with the National Travel Surveys to which they led.

The household survey of the London Traffic Survey of 1962 has already been discussed in the previous chapter. There were many other contemporaneous surveys: of freight and taxi movements, of cordon flows by both road and rail across the survey area boundary, and across internal screen lines, such as the river Thames, designed to give a complete picture of the movement of people and goods, together with a full description of the networks on which they moved.

It should be stressed, however, that the initial planning of the survey was done on a basis of road journeys only; the public transport networks were an afterthought. As stated in the final summary of the project (GLC, 1969b, para. A.42):

> 'When the London Traffic Survey was started in 1962, it was not envisaged that an analysis of public transport would be included in the project and consequently the terms of reference did not allow for the collection of satisfactory public transport trip data or for checks of the amount of traffic actually carried on the public transport system in order to calibrate and check the model.
>
> 'When it was decided to enlarge the scope of the study to include a more detailed analysis of public transport it was not practicable to collect much additional information on the subject. Available information indicated that the disparity between estimates of the number of trips made in London and those recorded in the home interview or postcard survey (of inbound rail trips) was not small enough to be ignored.
>
> 'It was decided to estimate an "unreported trip matrix" based upon the information available.'

The report continues with the procedures adopted to fill in this gap in the initial data.

This LTS household survey was the first such survey in the UK to be analysed by computer, and it is because of the way the computer works that the kind of analysis that I have been presenting in the previous chapters was not done. This statement requires some elaboration.

When I joined the GLC in 1966, my first job, as I have already noted, was to proof-read the second volume of the LTS, in which the structure of the transportation model was presented and used to describe the 1962 data, and to suggest areas in which I thought improvements could be made.

My first area of work was to tackle the household car ownership forecasting side of the model, which led to my first paper (Mogridge, 1967), the revised GLC car ownership forecasts (Mogridge and Eldridge, 1970), and ultimately to my first book *The Car Market* (Mogridge, 1983a). My interest in the inter-relation of car ownership and use has continued (Mogridge, 1989) and will be returned to in Chapter Nine.

The other area of work upon which I was soon embarked was to implement on the GLC's computers the transportation model package as used by the consultants in the LTS. Initially I named this the NETRAN (Network and Transportation Analysis) suite; after many transmogrifications since then, including the DTM (District Test Model), the STEM (Strategic Transportation Evaluation Model) suite, and the GLTS Model, it is now the model used by the Department of Transport via its current consultants as the LTS Model.

There were, and are, two essential problems with a strategic model covering an area the size of London.

First, all travel by individuals and all freight movements must be represented on the network. Now the processing time for the computer model goes up essentially as the square of the number of points representing the origins and destinations of travel. Some compromise is therefore needed between the fineness of detail and the cost of computation. In London, various levels of detail are used for various stages of the process, with the finest level being at the zonal, at which roughly 1,000 zones represent the origins and destinations of travel.

At this level of disaggregation, there are therefore 1,000,000 origin-destination pairs representing potential trips. Since in the household survey itself there are only about 250,000 trips recorded (only!), and most of those are short-distance, local trips, as we will see, about 95 percent of the potential trips do in fact have no surveyed trips to represent them.

It will be recognized that the finer the level of detail the more difficult it becomes to predict the number and types of households and all the other types of land use into the future, and consequently the number of cars owned and car parking places available.

It is, however, only with a fine level of detail that one can hope to obtain a good approximation to the access times and costs of public transport, as we have been discussing in the last chapter. The necessary compromize between the scale of the modelling and its cost has been one of the recurrent threads running through the history of the model.

Secondly, and much more importantly for the present work, the computer imposes a precision on the network which does not exist in real life. To be more specific, the computer requires that for every link in the network—whether road, bus route or rail service—a precise and unique time and cost of travel be provided.

It is only if such a precise and unique time or cost exists that a computer can calculate a unique set of routes between each pair of points in a network. As we have seen, however, each individual, when he or she sets out on a journey, has no precise knowledge of how long that particular journey will take on that occasion (although the cost may be well-defined).

The computer, then, has to work with the state of the traffic system which is in equilibrium, where average times and costs are well-defined and precise. The individual, however, on a particular journey will record journey times and costs which are specific to the occasion, and may well be far from average.

As we have seen, average speeds are indeed very stable, so that it is indeed possible to be very precise about the time and cost taken to traverse any link of a network at equilibrium. It cannot be done, however, by taking individual records from a household survey. And herein lies the rub.

One of the first attempts to overcome this problem was to take a random set of the actual journeys made in the LTS household survey and to re-survey them with a stop-watch to obtain precise times. This was done by Munt and Woodhall (1967) between May and August 1967. Some 240 journeys were selected, half from the car journeys recorded, and half from the public transport journeys recorded. The journeys were further divided into time periods corresponding to the morning peak, midday off-peak and evening peak, so that 40 journeys were chosen from each of these three periods by each of the two methods of travel.

The actual sample procedure was somewhat complicated since a number of surveys had to be used as a base to represent all travel in the LTS area properly. Using the equal interval method, the car-owning household sample was sampled at a rate of 1 in 380, whilst the non-car-owning household sample was sampled at a rate of 1 in 60 to compensate for the differing sample rates in the original surveys. Appropriate sample weighting factors were then allocated to each journey.

Journeys less than one mile in length were excluded, as were journeys not by car or public transport. Journeys were also excluded whose mid-point in time did not fall in the periods 7.45 am–9.15 am, 10 am–4 pm and 5 pm–6.30pm, corresponding to the morning peak, midday off-peak and evening peak respectively. External journeys were also excluded as these were covered by representative samples drawn from the other cordon surveys to cover road and peak-period rail journeys (non-peak rail journeys were not covered).

The final six sets of samples of 40 journeys originally made by car and public transport in each of the three time periods were then drawn from each consolidated list by an equal interval method, taking account of the weighting factors. The external journeys were only in-bound, so are given a further weight of 0.5 in the sample.

The journeys were then resurveyed both by the chosen method of travel and by the non-chosen, using reasonable routes (i.e. routes which were likely to be a minimum time route) and starting in principle at the same time as the original trip. It is not, however, clear from the original report what precise criteria were used for starting times. The car journeys were surveyed using an A.60 car, which is comparable to Bressey's Light Six car.

It appears that five minutes were added to car journey times to represent the access times of car journeys in the central area and one minute outside; the journey actually recorded was the kerb-to-kerb journey as in the random journeys described in Chapter Two. It is not stated whether this addition for access was based on the work of the TRRL described in Chapter Three (Inwood, 1966) or on any analysis of the LTS itself.

Table 5.1 gives the results found by Munt and Woodhall for these journeys.

These distances are measured along the road during the car journey. The conclusion with these results was, not surprisingly:

'It has been found that private transport is generally faster

Table 5.1 Comparison of journey speeds (all-day, LTS 1962)

	Within Central Area	*Central & Inner Area*	*Central & Outer Area*	*Central & External Area*
Number of journeys	6	26	31	7.5
Average distance (miles)	3.23	4.13	11.3	32.1
Average time (minutes)				
Car	25.7	25.5	48.2	118.0
Public transport	34.2	32.6	58.3	94.3
Average speed (miles/h)				
Car	7.54	9.72	14.1	16.3
Public transport	5.67	7.60	11.6	20.4
Ratio Car/PT speed	1.33	1.28	1.21	0.80
	Within Inner	*Inner & Outer*	*Within Outer*	*Outer & External*
Number of journeys	22	22	83	7
Average distance (miles)	3.50	6.81	4.95	8.96
Average time (minutes)				
Car	20.0	30.1	17.3	24.4
Public transport	30.0	46.0	36.7	50.3
Average speed (miles/h)				
Car	10.5	13.6	17.2	22
Public transport	7.00	8.89	8.09	10.7
Ratio Car/PT speed	1.50	1.53	2.12	2.06

than public transport for all journeys except the long-distance peak-period rail commuter services'.

As we can see, in outer London in particular, car travel is on average over twice as fast as public transport, mainly bus. In inner London, it is 50 percent faster, and it is only for the longer journeys to or from the central area that the car and public transport times approach each other.

Goodwin (1969) has analysed these results further, which allows us to make some comparison with the kerb-to-kerb road speeds of Table 2.4. He took the portion of the car journeys within the central, inner and outer areas respectively, during the peak-hours and in the tidal flow direction; that is, in-bound morning and out-bound evening flows. His results for the average running speeds are as follows:

Central area	11.2 miles/h
Inner area	14.6 miles/h
Outer area	19.6 miles/h

Whilst the Central result is slightly lower than the results in Table 2.4, the others are comparable, even though the sample size in Munt and Woodhall is low.

Goodwin also analysed the components of bus journeys, thus allowing a comparison with the random off-peak bus journeys in Table 3.4. His results are given in Table 5.2.These results are from a peak-period sub-set of the 86 journeys made by both car and bus, 43 of which were initially made by bus and 43 by car.

Table 5.2 Means of bus journey time components, peak-period 1967 (minutes)

	1 stage	*2 stage*
walk to first stop	5.4	4.0
walk to second stop		0.9
walk to end point	4.4	3.3
total walk	9.8	8.2

These are much lower values than in Table 3.4, partly because these are actual journeys not potential journeys. They also cover the whole area, not just the centre.

He also examined waiting times of peak-period bus journeys and found the following:

Central area	3.4 minutes
Inner area	4.5 minutes
Outer area	6.3 minutes

It is not clear from Goodwin's description whether this is the waiting time for each bus or for each journey, but it can be clearly seen that the Central area result is lower than the value in Table 3.1 for 1962 random off-peak journeys (4.7 minutes). This would seem to confirm that the bus waiting times were lower in the 1960s than now.

It is not at all clear from Munt and Woodhall's results whether any allowance had been made for the fact that the longer distance public transport journeys to the centre are usually made to a timetable; that is, the traveller knows when to leave home in order to catch the appropriate train.

The main problem with Munt and Woodhall's presentation of their results is that each of the two original methods of travel—car and public transport—and each of the three periods are given equal weight in the averages quoted above. Although rail trips to the centre in the morning peak and from the centre in the evening peak are by far the predominant method of travel, their results do not allow a proper appreciation of the journey times and speeds to be obtained.

At least, though, they did separate out the journeys originally made by car and those made by public transport, and show that the former had a higher ratio of car speed to public transport speed than the latter. This result is, however, partly due to the fact that the average public transport rail journey length is longer than that of car journeys.

It is perhaps surprising that no comparison was made with the actual times as recorded by the LTS respondents on whose journeys Munt and Woodhall's survey was based.

The conclusion of Munt and Woodhall's work thus appeared to be that car travel was always faster than public transport even within and to the centre, except for the longer-distance rail travel, whereas in practice this conclusion was suspect because of the procedures used.

The work was never repeated on subsequent surveys, no doubt because of Munt and Woodhall's conclusion that:

> 'It is a satisfactory but expensive way of getting point-to-point journey times'.

It is also rather surprising that no attempt was made at that time to examine those zone-to-zone pairs for which journeys had been made by both car and public transport in the LTS survey.

This has subsequently been done by Beardwood (1989) for the 1981 survey, and it is instructive to examine these results which are based on a uniform survey of households, rather than—as is the case with the 1962 survey—of separate surveys of car-owning and non-car-owning households with different sampling rates.

Although there are some 270,000 journeys in the 1981 survey, excepting those where members of the same household travel together, only some 7,500 journeys were by both private and public transport during the same time period; that is the morning peak, inter-peak, evening peak and rest of day. Most of these are intra-zonal, leaving 1,300 zone-to-zone pairs with at least one journey by both private and public transport; not surprisingly, most of these zone pairs are adjacent or local; that is, short-distance journeys.

If we examine the minimum time by private transport and compare it with the minimum time by public transport, we find that the private transport journeys are, on average, 20.1 minutes faster, with 91 percent of journeys being faster by private than by public transport. This is essentially because most of these pairings occur in outer London for short-distance journeys, where the alternatives are car and bus. This is equivalent to the result found by Munt and Woodhall in Table 5.1 Outer to Outer, where car times are 17.3 minutes and public transport (i.e. bus times) are 36.7 minutes.

If we take zone pairs with a destination in the Central Area, then the time difference is only 9.6 minutes, with some 29 percent of journeys having a faster public transport journey, as opposed to 9 percent overall. Again, this is very similar to Munt and Woodhall's result for journeys to the Central Area.

Munt and Woodhall's attempt, therefore, to obtain accurate point-to-point journey times reinforced the prevailing viewpoint that car travel was always faster than rail or bus travel, except in the exceptional circumstance of the long-distance commuter. Two main consequences followed.

First, no attempt was made to calibrate the transportation model against the actual recorded journey times in the surveys. They were too imprecise (as we saw in Figure 4.1) and not unique. Indeed, it appears that only off-peak journey speeds were available, as we have already seen in Figure 4.6. No attempt was reported even to reproduce these time contours from central London.

Instead, calibration of the model allocation of trips between zone pairs and between methods of travel was done against the calculated journey time from the network, which was precise, although only in so far as the zonal centroid could be said to represent the travel origin or destination.

Secondly, and more importantly, the transportation model was split into a car travel sub-model and a public transport travel sub-model. Whilst this was and is standard practice—and will no doubt remain so—the crucial interaction between the choice of car and of public transport for a particular journey was expressed by a particular mathematical formulation which could be calibrated against the existing data but could not be verified by it. This interaction is expressed by the modal split sub-model, where a mode is a method of travel, either car (private) or public transport.

Calibration is where one matches the model to the existing data. Obviously the simpler the model the worse the match. The aim is to match the data with as simple a mathematical representation as possible which still represents all that is thought to be relevant about travel behaviour.

Verification is where one matches the model to another set of data collected at another time or in another place to confirm that the relationships that one has put in the model are in fact able to be used for prediction. At the very least, one requires stability over time if not over place.

The only London household surveys available for the verification of the structure of the LTS model were the London Travel Surveys of 1949 and 1954 already mentioned. But whilst the LTS Volume 1 report mentions the latter survey (paragraphs 6.60–6.62 and Tables 6.18–6.19), this is merely to compare the distribution of the modal shares of the various journeys, and especially by the purpose of the journey, in 1954 and 1962. This comparison showed the growth of car use, and especially of the use of the car for non-work journeys, and the decline of bus use.

The LTS Volume 2 report makes no mention of the earlier household surveys in its verification procedures, but merely mentions the total journey trends as measured by traffic statistics.

Thus the LTS model relied heavily on comparisons from the United States for its verification of the structure of the model, thus using spatial comparisons rather than temporal. The modal split sub-model was, in my opinion now, wrongly specified, but this could not be verified at that time. Calibration was adequately specified however, but the calibration was on the calculated trip times rather than the reported trip times as already noted, thus hiding the crucial part played by access times (or rather a composite of time, cost and effort) in the modal split process in congested conditions.

It is worth reflecting on why this mis-specification in the modal-split model should have occurred, since it is so important for the way that transport policy has developed since then. Whilst the results of Munt and Woodhall reinforced the then current opinion that cars were always faster for particular journeys, the correct view had been available since at least the work of the economists Pigou (1920) and Knight (1924), which we will discuss in the next chapter. It had, however, been obscured by the engineering criteria expressed by Wardrop (1952) on which all transport network models are based.

Wardrop (based on work since the establishment of the Road Research Laboratory in 1933, and especially in the immediate post-war years) stated two criteria as follows:

> '1) All routes chosen by a driver between two given points in a network will have the same trip time and all routes not chosen will have an equal or greater trip time.
> '2) The travel time in a network will be a minimum'.

It will be noted that these criteria are strictly for car drivers and refer to travel times only. The rationale for this was that all car drivers experience approximately the same costs and effort in travelling, so that one could therefore ignore cost and effort and concentrate solely on time.

This is a reasonable supposition for car drivers, but it had the unfortunate effect of obscuring the economic criteria which could be expressed in a completely analogous manner, namely:

> 1) All routes and modes chosen by a traveller between two given points in a set of networks will have the same journey cost, where cost is a composite measure of trip time, money and effort, and all routes not chosen will have an equal or higher cost.
> 2) The travel cost in a set of networks will be a minimum.

It is usual to call this measure of composite cost a generalized cost, to differentiate it from one measured solely in time or money. It requires for its evaluation a measure of equivalence between time, money and effort. This is usually expressed as the value in money of time or effort; that is, how much a traveller is willing to pay to save time or effort. The first modern measure of the value of time was by Quarmby (1967), using a survey of commuters in Leicester, although—as we saw—it had been anticipated by Horne in 1846.

Two features of the transportation model acted against the recognition of the validity of these latter criteria: the separation of the sub-models for cars and public transport, and the use of zonal centroids to represent the origins and destinations of trips in a zone. For a journey between any given zonal pair, one had a private transport cost and a public transport cost which were based on the zonal centroids and the minimum cost path through the respective network found by the computer, not on the actual costs of the journeys as made by any actual travellers. There are, in fact, relatively very few zonal pairs for which journeys by both private and public transport are sampled, except for the very short journeys of a mile or so, as we have noted above. Even with such zonal pairs, because a zonal centroid is not the actual origin or destination of a trip, the costs of any traveller making that particular journey between a given zone pair may vary substantially according to his or her precise problems of access to the networks, even if the journey costs on the networks are correctly represented by the minimum paths.

These are two good reasons why Munt and Woodhall chose to survey journeys both by the chosen mode and by a reasonable alternative using the precise addresses of the origin and destination.

Given the spread of costs that locating journey ends by a zonal centroid causes for a given journey between a given zone pair on a given network, it is understandable that a modal split model which calculated the proportion of journeys made by each mode between each zone pair according to the calculated minimum costs on each mode seemed quite a reasonable model. This assumes that some travellers between a given zone pair will find one mode has a lower cost, whilst others having different trip ends in the same zones will find the other mode has a lower minimum cost. The criteria for an equilibrium to be achieved in the traffic pattern had been set out by Beckman, McGuire and Winsten (1956). Essentially, two requirements had to be met.

1) The costs on a given link had to increase, or at least not to decrease, with the flow on the link.
2) The demand for travel between two points had to fall as the costs of travel between the points increased.

Given these conditions, the flows on each link could be uniquely determined and the travel pattern calculated, though not necessarily the precise routing from each origin to each destination. Many different routing patterns might combine to give the unique flows on each link. The importance of these two criteria was again obscured by the way in which the private and public transport models were split into two, with the trips on each network being calculated separately.

In Phase II of the LTS, the assignment of the road traffic to the road network was done on the basis that existing off-peak road speeds would continue to be applicable in the future, and that the new roads proposed would operate at their design speeds. A computer procedure based on Wardrop's first criterion could then be used to find the minimum time path from any origin to any destination, and all traffic between that zone pair would then be assigned to the path found. This was the then standard United States Bureau of Public Roads procedure.

This gave some peculiar results in the assigned flows. For example, the highest flow produced for the 1981 target date was on the West Cross Route, an inner London orbital motorway. The flow was 339,000 vehicles per day, which would require a 10-lane dual carriageway at the design speed!

In paragraph 17.38 of the consultants' report, we find the following apologia:

> 'The volume assigned to the West Cross Route . . . greatly exceeds the average weekday traffic volume on even the busiest urban freeway in the United States of America. However, assigned volumes of this order of magnitude are by no means peculiar to London. A study completed in 1962 produced an assigned volume of over 500,000 vehicles per day on one

section of a motorway penetrating the central area of Boston, and future volumes of over 200,000 vehicles per day were assigned to sections of the motorway systems tested for Detroit, St. Louis, Washington and Kansas City.'

Paragraph 17.39 continued:

'For comparison, some of the highest average daily traffic volumes recorded on facilities in the United States of America are the following:

Harbor Freeway, Los Angeles	200,000 vpd
Dan Ryan Expressway, Chicago	175,000 vpd
John C. Lodge Expressway, Detroit	165,000 vpd
James Lick Freeway, San Francisco	160,000 vpd
Schuylkill Expressway, Philadelphia	140,000 vpd
San Francisco/Oakland Bay Bridge	125,000 vpd
George Washington Bridge, New York	100,000 vpd'

The implication is 'Don't worry, chaps! It won't be as bad as all that'. Interestingly, no assignment of the 1962 calibrated model trip ends to the 1962 network was presented in Chapter 17, merely the 1962 surveyed trip ends.

The output of Phase II, however, was thus not an equilibrium assignment to the road network, because only an all-or-nothing assignment was made. The problems this gave rise to were examined in Reid and Cottee (1967), for which I provided the computer programming.

In order to attempt to avoid the worst problems caused by the reliance on such an all-or-nothing assignment, I developed a computer procedure I called SCAMP (Shorter curvi-linear alternative minimum paths) (Mogridge, 1970b). This examined the all-or-nothing assignment from both ends of the journey simultaneously. It then calculated alternatives to the minimum time route by examining the central region between the two ends of the journey and finding routes that were distinct, but still minimum paths from each end, through the central region, and that were no more than a certain specified time longer than the minimum.

Figure 5.1 shows the alternative routes produced by the procedure for a journey right across the London region from the North-

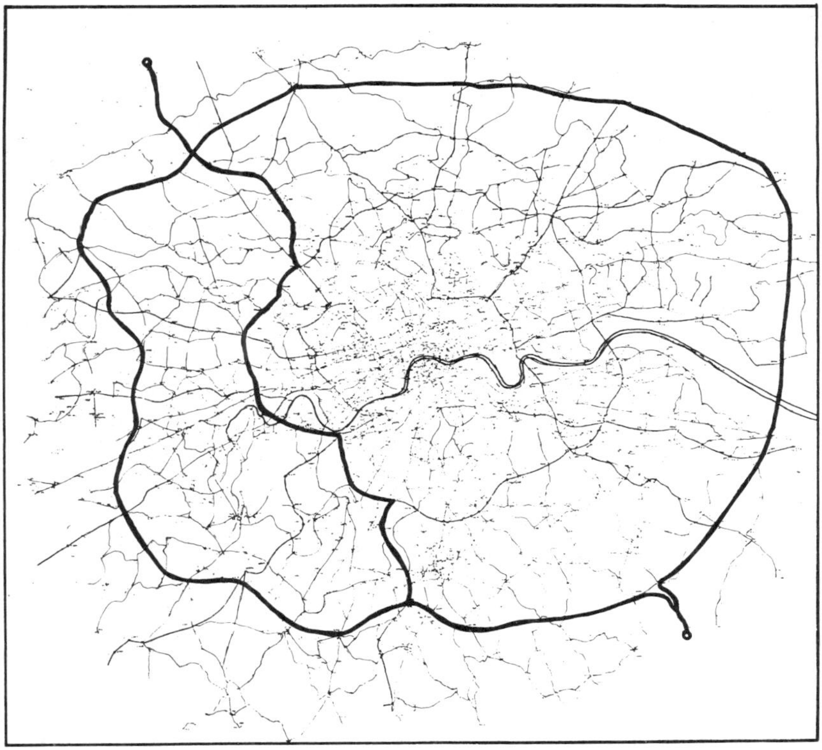

Figure 5.1 Calculated alternative minimum routes by the SCAMP procedure

West to the South-East on one of the planned networks. It can be seen that routes round the North Circular Road, and round the outer ring-road both clockwise and anti-clockwise, satisfy the criteria specified, but not right through the centre.

It will be recalled that Bressey and Lutyens, in their 1936 survey, had measured routes through the centre of London and circumferentially along the North Circular Road. They concluded:

> 'A motorist can, therefore, expect to make the entire northern circuit of London, via the North Circular Road (22.75 miles) in somewhat less time than he would spend in painfully threading his way along the direct and congested route (12.6 miles) through the heart of London'.
>
> Bressey and Lutyens (1938, Paragraph 25)

A simple calculation of the ratio of the lengths of the semi-circumference of a circle compared to its diameter shows that speeds on such a simple example of alternative routes must be in the ratio 1.57:1 for the two routes to have equal time. The ratio of the Bressey and Lutyens speeds (23.6/12.5) is 1.89, so the orbital is faster.

This particular example of alternative routes continues to fascinate people. On 25 September 1988, the TV programme *Equinox* for Channel 4 presented a 'Road Test' whereby drivers were sent out at 8.30 am from Milton Keynes (to the North-West of London along the M1) to Maidstone (to the South-East of London along the M2) on equivalent routes, right through the centre and both clockwise and anti-clockwise round the M25, the outer London orbital opened in 1986. Whilst the M1 and M2 are not strictly opposite each other on the M25, it is instructive to compare the times.

Clockwise M25	107 miles	137 mins	47 miles/h
Anticlockwise M25	119 miles	152 mins	47 miles/h
Central	92 miles	175 mins	32 miles/h

The central route includes a substantial proportion of motorway driving to and from the M25, so the speed is much higher than earlier values given across London. It can be seen that the M25 provides a faster route than going through the centre, but not by much. The rush-hour was particularly bad on that day, with a number of traffic jams.

The programme took the opportunity of comparing the time by the recently re-opened Thameslink service from Bedford, via changes at King's Cross and Bromley South, to Maidstone, to demonstrate the use of the cross-London rail link and to enable me as the participant to express the arguments which I will be developing later in this book.

The recorded journey was as follows:

Thameslink	96 miles	163 mins	35 miles/h

It can be seen that this is similar to the journey through the centre.

The GLC traffic engineers used my computer procedures to assess the all-or-nothing routes provided by the minimum time path, and to change them to an alternative where the engineers thought the minimum was an unrealistic path.

The first major use of my work on the transport model at the GLC in 1967 was in fact to assign the estimated 1981 car trip ends, produced by the consultants in the LTS Phase II, to the Greater London Development Plan (GLDP) motorway networks, in order to aid in the assessment of the proposals. It should be remembered that at this stage (Phase II of the LTS), the motorway network was assumed to be able to operate at its designed speeds, which varied according to location. Even with a fixed speed on the motorway network, this did not contravene the first criterion of Beckman *et al.*.

In fact, the consultants recognized the shortcomings of this approach as they stated in the Phase III report:

> 'Although the methods of trip prediction, distribution and assignment used in Phase II of the Study provided a useful basis for the planning of a future road system in London, they had several shortcomings. In particular they were insensitive to the quality of public transport for a particular journey (public transport trips were neither distributed nor assigned), they did not take into account the effects of congestion on particular roads, and, perhaps most important, they did not relate directly the amount of travel predicted to the capacity of the transport system available. The model used in the present phase has been designed in an attempt to overcome these deficiencies.'
>
> LTS (1968, Chapter 20.1)

The model used in Phase II had already had to reduce car trips substantially in the calibration process, to limit car trips estimated in 1962 to the values of car trips actually observed in particular areas, particularly the centre. A simple, but arbitrary, linear programming technique was used to reduce trips *pro rata*.

In order to do the assessment of the motorway proposals with the estimated traffic flows produced by the consultants in Phase II in 1967, we had, in the GLC, to develop a method of limiting the access of short-distance trips on to the motorway network, otherwise the motorways were swamped by such short trips as just noted.

Thus flow had to be limited to the capacity at which the motorway would operate at its designed speed.

This was a simple enough computer procedure; it was merely necessary to specify a hierarchy of the network, and then one simply forbad short-distance trips from access to the highest, motorway level. Whether it was possible in real life to prevent such trips accessing the motorways was quite another matter.

Considerable development of these modelling techniques took place after I left the GLC in 1968, initially as the District Test Model (DTM), and using the output trip matrices produced by Phase III of the LTS. The results were discussed at great length in the inquiry into the GLDP in the period 1969–71. Unfortunately, the computer procedures used in the hierarchical evaluation, and the comparison with the DTM procedure which made no such distinction in the hierarchy of roads, were never written up formally, essentially because it was found that there was in fact little difference with the latter, more general procedure.

The fact that it was possible to consider working in this way was, no doubt, in part due to the split in the way that private and public transport was assessed.

In Phase III of the LTS the development of the modelling procedures included at last a public transport assignment model, described in Chapter 20.26 of the LTS report (LTS, 1968) and in Appendix A of the summary report by the GLC *Movement in London* (GLC, 1969b), in which a single network for bus and rail was formed.

This was initially a technique developed by the USA Bureau of Public Roads, but later work used a procedure developed specifically for London by the consultants, called TRANSITNET. This was operated at the 1,000-zone level, unlike the road network, which was operated at the 250-district level. The rationale for this was quite simple; buses had to be assessed at zonal level, as trip lengths by bus were too short to be adequately modelled at the district level, whereas trip lengths by car were long enough to be adequately modelled at the district level, which reduced computer costs.

The travel time for a public transport trip was the lesser of the minimum rail time and the bus time, the latter being the off-peak road time multiplied by a factor of 1.8 plus a further 10 minutes representing access (GLC, 1969b, Paragraph A.16). The trip time for a given trip was thus constant, irrespective of flow, just like the road network in Phase II.

The most significant change to the model in Phase III was, however, the 'constrained demand' model. The consultants had moved on from constant road speeds on the motorways to incorporate a 'congestion' effect, as just noted.

They had also devised some quite arbitrary procedures in order to limit forecast car trips, particularly to the centre. The capacity of the transport system was taken into account by a linear programming technique to reduce the trip generation of car owners in particular areas, re-routeing some trips, diverting some trips to public transport and suppressing other trips.

There were two parts to this process. Motorway speeds were reduced by a procedure which left the motorways as equally overloaded as the rest of the road system, using iterations of redistribution and reassignment with the changed trip times.

The total number of car trips was then reduced by a procedure which left the total road system carrying the maximum amount of traffic within its capacity. Trips were either diverted or suppressed in each area by an amount depending on the ratio of car driver to public transport trip rates in car-owning to non-car-owning households for three trip purposes (work, other-home-based or non-home-based trips).

That is, if there was not enough capacity on the road network to take all the projected car trips, then some car-owning households would be assumed to behave like non-car-owning households.

The second part of Beckman *et al*'s criteria was that trip numbers should fall as trip costs increased. In the distribution and modal split models of Phase III, this criterion appears to be fulfilled, in that functions are defined which have precisely this effect.

There is, however, a proviso. The functions are defined from the attraction end of the trip (i.e. the workplace, the shopping centre and so on) and not from the generation end; that is, the home etc. This overcomes a rather curious effect, that trips to the centre of London are much longer than trips to other destinations in the suburbs; that is, that people are prepared to travel further to the centre of London than they are to places in the suburbs, such as to work or to shop.

Obviously, one can make suppositions as to why this should be so. Wages are higher in the centre; there are more specialist shops and services such as financial, legal and medical services. People are thus prepared to travel further because the benefit is higher.

Nevertheless, strictly speaking, Beckman *et al*'s second criterion is not met by trips with destinations in the centre. One can only satisfy the criterion if one takes trips as if they had origins in the centre; that is, an attraction there, as was actually done in the Phase III model.

The actual form of the distribution and modal split model used was extremely simple, based on journey time as assessed by the minimum path calculations of the assignment, but the models were given separate parameters for every district (of which there were about 250).

It is worth noting at this point that the car trip times from the assignment had three minutes added to them to represent parking, except in the central area where a further three minutes were added (GLC, 1969b, Paragraph A.17). It is not stated on what basis this time was arrived at.

The Phase III model was used to test a number of combinations of road and public transport plans for the projected 1981 demand. These were selected from an extensive set of plans drawn up by the consultants in consultation with the various authorities and operators.

The three road plans tested varied from a do-minimum road plan at one extreme to a very extensive set of motorway proposals which could cope with the entire demand for car travel at the other, with an interpolated medium set of motorway proposals in between.

The two public transport plans tested varied from an extensive programme of improving rail services together with some new rail lines and considerable improvements to the bus network, to a plan which also included a cross-rail scheme connecting Paddington to Liverpool Street and Victoria to London Bridge.

Paragraph 4.36 of the summary report (GLC, 1969b) states first of all:

> 'Prior to restraint there were virtually the same number of internal driver trips for the three road tests and the same number of public transport trips from the modal choice calculation'.

It goes on to describe the level of restraint which these tests required in order to balance the demand with the capacity. Whereas the maximum motorway plan could cope with all the demand, the medium motorway plan required a restraint of 7 percent of road trips, of which 40 percent were transferred to public transport and 60 percent were suppressed. The do-minimum road plan required a restraint of 27 percent of road trips, of which approximately the same proportions transferred and were suppressed as in the medium test.

Interestingly and inevitably because of the techniques used, paragraph 4.37 goes on to note that the level of public transport usage did not vary between the different road tests except for the transferred trips from the restraint procedure.

More significantly, paragraph 4.38 reveals that the cross-rail scheme had no effect on total passenger trips on the public transport system, although it did divert passengers between the public modes and presumably between destinations. There were, therefore, no additional passengers on the rail system transferring from cars because of the additional rail facilities. This again was inevitable because of the structure of the modelling process.

At this stage of transportation modelling, therefore, there were, despite the enormous progress in the ability to describe and predict future travel patterns, some very severe shortcomings in the techniques.

A good introduction to the techniques of transportation modelling at that time is the book prepared by my colleagues in the course we gave at the Polytechnic of Central London, in which I ran the computer workshop (Lane, Powell and Prestwood Smith, 1971). This gives the mathematics of the various processes in more detail.

The GLC noted (GLC, 1969b, paragraphs 5.8–5.14) that:

> 'The forecasting of future levels of car ownership is crucial to the planning of transport systems. The possession of a car affects both the use of the road system and the use of the public transport system and has a considerable number of side effects such as the parking demand at both ends of a trip and impact on the environment.
>
> 'Work done by the GLC (ie by Mogridge, 1967) has attempted to carry these (RRL) earlier trend studies (of the

growth of national car ownership) further . . . A much higher saturation level of car ownership than that forecast by the RRL is predicted . . . but the difference between the RRL and the GLC forecasts in 1981 is not large. In the years following 1981 the difference becomes significant.'

The GLC area forecasts of car ownership were reported in detail in Mogridge and Eldridge (1970), as already noted.

The key issue that the consultants had raised was quite clear. If it was impossible for social and environmental reasons to build the motorway network which would enable all the unrestrained demand for car travel to be met in 1981 (the forecast year), was there a method of restraint that was better than the restraint by congestion which would otherwise occur? As car ownership continued to grow beyond 1981, this question would become of increasing importance.

One of the decisive factors in forming public opinion about the motorway proposals was undoubtedly Thomson's book *Motorways in London* (1969). This was a report by a Working Party chaired by Thomson and set up by the London Amenity and Transport Association, which was formed in response to the motorway proposals. He criticized the motorway programme on three main counts.

1) Inadequate consideration had been given to the provision of the access and egress that traffic using the motorways would require on the local road system;
2) The motorways would in themselves generate a lot of extra traffic that would not otherwise have existed; and
3) The motorways would draw off traffic from the public transport systems which would as a consequence deteriorate.

It can be seen that Thomson's first point was essentially one aspect of the problem which I have been describing, and points two and three were effectively ignored in that assessment, as Thomson's challenge maintained.

Thomson's second point was essentially the same as that raised by Downs (1962) in his *Law of Peak-Hour Expressway Congestion*, which he likened to Parkinson's (1960) Second Law adapted to traffic: 'On urban commuter expressways, peak-hour traffic congestion rises to meet maximum capacity'. Downs discussed the cases of a new expressway when the alternatives available were the local street system, a bus system which could be adapted to use the new expressway, and a rail system.

I will return to the discussion of public transport alternatives later in Chapter Seven.

Thomson (1969) also argued that the railways acted as a 'cushion' to prevent congestion reaching unacceptable levels. I will return to this point too in Chapter Seven.

I note at this point Down's argument about the establishment of a new traffic equilibrium by route shifting. He first noted that traffic speed decreases as flow increases, using data from Watson, Smith and Hurd (1955) as his authority. This is reproduced in Table 5.3.

Table 5.3 Speed/flow relationship for expressways of four or more lanes

Vehicles per lane per hour (maximum)	*Speed (miles/h)*
1,500	35–40
1,250	40–45
1,000	45–50

Watson, Smith and Hurd (1955, p.382)

Downs continued:

> 'The expressway is shorter than the other routes; hence motorists traveling at the same average speeds over all routes will still arrive at their destinations faster on the expressway. Therefore route shifting on to the expressway will continue until the average speed on the expressway is reduced to *below* the average speed on alternative routes.
>
> 'Clearly, the expressway was designed to handle traffic at *higher* speeds than previously existing roadways; yet, at equilibrium, traffic on the expressway is moving *more slowly* than traffic on the older routes. Therefore we can say that *congestion on the expressway has risen to surpass its optimal capacity*. This result is a necessary outcome of the forces of traffic equilibrium.' Downs (1962, p.397)

This point was taken up by Leavitt (1970) in her book *Superhighway —superhoax*. She gives a number of examples (pp.258–64) of expressways in urban areas where the traffic flow exceeded their design capacity within a few years of opening.

Downs goes on:

> 'The only way in which such heavy congestion could be avoided is through creation of a single expressway or system of expressways with such an enormous capacity that all the commuters formerly driving on parallel conventional streets could use the new facilities and still maintain optimal speeds.'
>
> Downs (1962, p.398)

This is essentially the same point as that raised by the LTS Phase III 'no restraint' maximum motorway plan.

It is interesting to recall that, in summing up the value of their Highway Development Plan in the film *The City* (1939), Bressey had argued at the end of the film that the creation of the proposed system of highways for London would enable him to get home from his office not in the one hour it currently took him, but in only 15 minutes!

In fact, as we saw earlier, despite all the hundreds of millions of pounds spent on upgrading the North Circular Road to a dual-carriageway, grade-separated highway, the speed of travel remained exactly the same in 1986 as in 1936, amply demonstrating the effect of Downs' Law. The M25 orbital motorway around London, opened in 1986 and subsequently already widened to a four-lane carriageway in the South-West section, shows the effect even more vividly.

The public's view of the motorway proposals in the 1969 GLDP was able to be expressed in the GLC election of 1973, when the issue became one of the determining factors in the election, with a 'Homes before Roads' slogan, coupled with other concerns about the effect on other land uses and on the environment. The proposals were decisively rejected by the electorate, and the in-coming Labour administration lost no time in scrapping them.

Thomson recognized that the data on which such challenges were based was in fact inadequate to enable improvements to the relevant areas of the transport model to be made at that time; that is, prewriting of the book published in 1969. He did, however, draw on the Royal Commission on London Traffic of 1906 and the Highway Development Survey (Greater London) 1937 (Bressey and Lutyens, 1938) to illustrate his thesis, as we have done in earlier chapters.

This rejection of the motorway proposals and the adoption of a policy of traffic restraint raises the whole issue of road pricing to which we will now turn, as it was to occupy much of the debate about transportation modelling in London for the next decade.

A number of advances in transportation modelling had to be made before the question could even be tackled, but first we look at the history of road pricing as it was up to 1969.

6 *Road Pricing*

In the path-breaking paper by Smeed (1961) there is a clear statement about the impossibility of satisfying all demand for car travel in cities above a certain size and with an employment concentration at the centre of the city of a certain level. Smeed achieved this by a simple mathematical treatment in which he considered only the journey to work to central places of employment.

All employment in the centre was assumed to be at a uniform density within a circle of a given size. All residences of workers were assumed to be at a uniform density, outside the central employment circle, and contained within a larger circle. All journeys to work within the residential area were radial, and all journeys from the central area boundary to a central employment place were randomly distributed and in straight lines across the centre.

With these simple assumptions, together with some about road traffic space requirements, he showed how it was impossible to provide enough road space for every worker to travel by car to work in the central area within an acceptable time period once certain limits of size and density had been exceeded.

He went on to consider the space requirements of parking, both at street level and multi-storey, and of the various forms of public transport, making assumptions about the space requirements per passenger moved.

He concluded that, beyond a certain size and concentration of city activities, cars could not be used to provide for all movement by individuals. They were too inefficient in their use of space, and some more efficient method of using space was required to move people if the size and concentration of the city was to be maintained.

In Smeed (1968) he amplified this point by stating quite clearly that the workers who could not travel by car would have to go by public transport, or change their home or workplace, or stay at home.

Thomson (1977) in his *Great Cities and Their Traffic* gives some estimates of the traffic intensity in city centres around the world which lends support to Smeed's theoretical calculations. Thomson took the size of the city centre as defined by the city authorities, approximated it by a circle of the same area, and took the average distance travelled by workers within the city centre to be equal to the radius of the circle so defined. (This is 13 percent less than the distance travelled within the central area using Smeed's assumption of straight lines, and 36 percent less than using a route factor of 1.20.) The travel intensity is then this average distance travelled, multiplied by the number of workers in the city centre and divided by the central area of the city. Thomson is able to quote from 22 cities using data from the period from 1963 to 1973. His Figure 2 is reproduced here as Figure 6.1.

To quote Thomson (1977, p.67):

> 'It appears that when the number of jobs in the city centre exceeds 50,000 to 100,000, the use of cars for commuting normally becomes limited by lack of space. This is only an

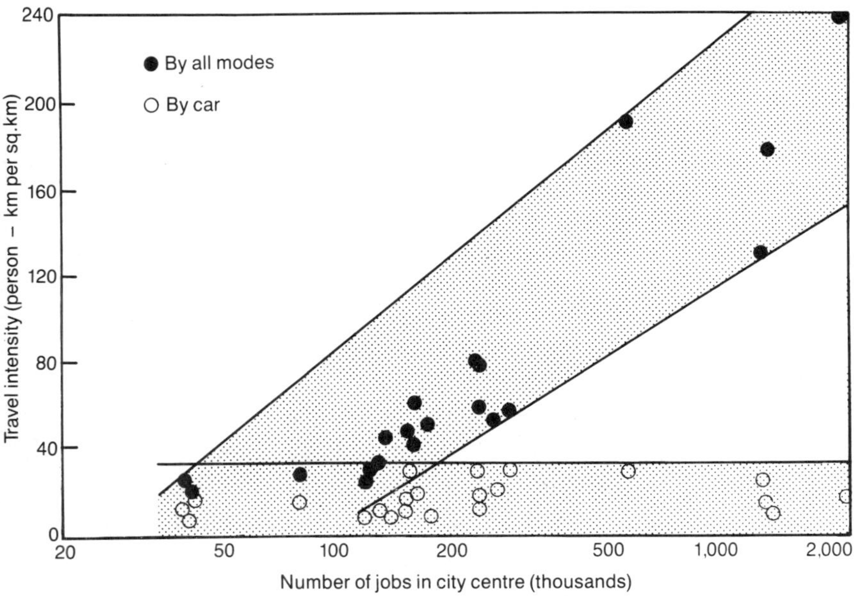

Figure 6.1 Travel intensity of work journeys in city centres

> observation, not in any sense a law, because doubtless it would be physically possible, by multi-deck methods, to design a city centre for a million employees in which everyone could commute by car.'

Whilst Thomson was gathering this data, it so happened that our paths crossed in Hong Kong, while I was working there in 1971 on an assessment of the proposed cross-harbour road tunnel (Mogridge, 1971). Unfortunately, he arrived just before a powerful but intense typhoon. This marooned me on Hong Kong Island, while he was stuck in Kowloon on the mainland, so our only communication then was telephonic. Although Hong Kong ('fragrant harbour') was chosen as a harbour site for its relative safety during typhoons, this did not prevent a number of large ocean-going ships from being wrecked and over a hundred sailors drowned.

Typhoons of such an intense nature play havoc with normal traffic patterns. The local paper had a photograph afterwards of multi-level parking under these conditions; cars washed down a steep hillside to land in a heap on top of one another at the bottom. Not quite what Thomson had in mind for multi-level cities!

The redesign of city centres to allow for full motorization had already been considered in some detail in the Buchanan report (1963). The authors realized that road capacity is, and will always be, below potential demand in places like central London. Using some simple trip length distributions for car journeys, they investigated the possibility of designing from scratch an area of central London for the motor car and concluded in paragraph 335:

> 'We had satisfied ourselves that even given complete and total redevelopment of a central high-density site, there would be a strict limit to the amount of traffic it could accommodate, dictated not so much by its own capacity as by the feasibility of providing a network to serve it and all the adjoining areas. Even so it was evident that to accommodate the maximum flows from the network (which satisfied all essential needs) would require a highly intricate multi-level design.'

In their conclusions (paragraph 457) the authors stated:

'In the long run, the most potent factor in maintaining a 'ceiling' on private car traffic in busy areas is likely to be the provision of good, cheap public transport, coupled with the public's understanding of the position. This last is essential, and is indeed one of the reasons why we have endeavoured to write this Report in non-technical language for the general reader.

'But the attractions of the private car are very great, and there can be no denying the difficulties of providing public transport services so intrinsically *convenient* that they will attract optional car traffic off the roads in appreciable quantities. But, given a different financial policy, travel by public transport could be made *relatively cheap*, and this may prove to be the key to the problem in the long term.'

In the Steering Group Report (Crowther) accompanying Buchanan's report, this was amplified further as follows.

'Regarded not as a solution in its own right, but as one arm of a coordinated policy, we think the case for expanded public transport in cities is proved. Without questioning, in its general application, the policy of making transport pay its way, we think the particular case of urban passenger transport needs to be considered in a wider context than the simple comparison between what can be collected in fares and what the service costs to run.

'Nor does this involve bringing in the inevitably vague conception of "social requirements". It ought to be a strictly economic calculation. In any given city there is a calculable number of bodies to be moved between home and work and back again every day. The number that can possibly be carried in private cars, even after an extensive programme of road-building, is also calculable and limited.

'The remainder will necessarily have to use public transport, and the means of providing it is one of the essential elements of the sort of coordinated and comprehensive planning that is clearly needed.'

(paragraph 27)

The last sentence could be a paraphrase of an identical sentiment expressed in the influential Brookings Institution study by Owen (1956).

Buchanan was right to emphasize that 'the public's understanding of the position' was a necessary factor in devising a coordinated and comprehensive plan for movement.

Despite the rejection of the motorway plans in London in 1973, there remained and remains a certain part of the public who are convinced that there must be a way in which enough roads can be built to satisfy the pent-up demand for car travel.

Despite, too, the failure of the subsidization of public transport fares, particularly in the period 1973–75, to attract sufficient car drivers off the roads to improve the overall traffic speeds, there remained and remains a certain part of the public who are convinced that if only public transport were made cheap enough it would solve the traffic problem.

These two opposing strands of policy—build more roads or subsidize fares—continued throughout the period from 1969 to the present to dominate political argument. One particular manifestation of this was the GLC's production of a long-term strategy (GLC, 1977) based on a non-network model which I will discuss in the next chapter.

The technical argument about the network model, however, was dominated by the need to appraise realistically the effect of price upon travel, and in particular to be able to assess a road pricing policy.

Road pricing in the UK has a very long history indeed. Whenever a royal charter was issued to allow a toll to be charged by the owner of a facility (such as a bridge or a ferry or a private right of way) in order to maintain its public operation, we have essentially a road price. Some still exist, such as at Dulwich in inner south London.

The toll system was extended to public roads themselves in 1678, when the deplorable state of many of the highways, which up to then were the responsibility of the parishes through which they ran, led to the Turnpike Acts. A turnpike was a barrier made of pikes or spears erected across a road to prevent passage, initially of attacking horsemen; the turnpikes at the toll-gates gave their name to the roads.

It is not surprising, therefore, that one should be able to find early discussions of the appropriate toll to charge for the use of such a facility. The foundations of the subject of road pricing were laid by a French engineer, Dupuit (1844), in discussing the benefit to society from the imposition of a toll on a footbridge. The paper is translated in Barback (1952) and I follow Goodwin and Jones's (1989) treatment.

Goodwin and Jones argue that Dupuit's presentation is a simple proof that the value, or utility as an economist would put it, to society of a bridge is always more than the money which can be collected from the tolls on its users. The presentation has not been bettered by later and more complex treatment, so I, like Goodwin and Jones, repeat a section of it here:

> 'We want to know the utility of a footbridge which is being used free of charge at the rate of 2,080,000 crossings annually. Suppose that a toll of 0fr.01 would reduce the number by 330,000, that a toll of 0fr.02 reduces it by (another) 294,000, and so on. We then say that for 330,000 crossings, the utility is about 0fr.01 and that for the next 294,000 crossings the utility is about 0fr.02 . . .'

Dupuit then draws up a table of the utility that would be obtained at every possible level of charge from zero to that price (0fr.15) at which there is no demand, and continues:

> 'Thus 102,000 francs would be the absolute utility to society of the bridge. We can find the relative utility by deducting the costs of maintenance and the interest on the capital expended in construction. If this latter sum were to reach or exceed 102,000 francs, the construction would have produced no utility, the difference expressing the loss which would have been made. Such is the calculation to be made when the crossing is free of charge.
>
> 'If there is a toll we must take only the figures below that of the charge. Thus for a toll of 0.05fr, for example, the absolute utility of the bridge is expressed by the sum of the ten last figures, or 66,000 francs; the utility lost, by the sum of the first five, or 36,000 francs; the product of the toll would be 770,000 crossings at 0fr.05 or 38,500 francs. With this toll, then, the

possible utility of the bridge would be distributed in the following manner:

To the toll collector	38,500 fr
Derived by those crossing the bridge (66,000–38,500)	27,500 fr
Loss of utility arising from the 1,310,000 crossings which would have been made but for the toll	36,000 fr
Total	102,000 fr

'As the toll increases, so does the utility of the bridge diminish in proportion; it becomes zero when the toll equals 0fr.15, at which price no one crosses the bridge; it is therefore possible for the loss of utility to rise to as much as 102,000 francs.'

Goodwin and Jones continue:

'There are three points of importance to policy analysis in Dupuit's treatment:

(1) The utility to users is measured by reference to what they are prepared to pay (but is not equal to what they do pay).

(2) There is a loss of utility incurred in charging a toll, derived from the crossings which are deterred.

(3) The utility gained by the "toll collector" must be included.

'These three points all focus attention on the inherent, inevitable, distributional aspects of pricing. Dupuit was well aware of indirect and distributional consequences.

'Overall, the argument tends to justify lower, not higher, charges. These are not necessarily zero: Dupuit provides the need for . . .

"A revenue sufficient to cover the cost of upkeep and interest on capital."

but tolls should be no more than this, ie in general they should be less than they would be if set at a profit-maximising level.

Almost the entire framework of the economic argument for road pricing and—implicitly—the likely political sensitivities are contained in this argument. The one crucial exception is the issue of congestion, which was introduced later.'

It should be obvious that there is a toll rate which raises the maximum toll revenue. Revenue is of course zero when the toll rate is zero, and when the toll rate is so high that users are zero, in Dupuit's case at 0fr.15.

This can be easily seen by considering the demand function shown in Figure 6.2.

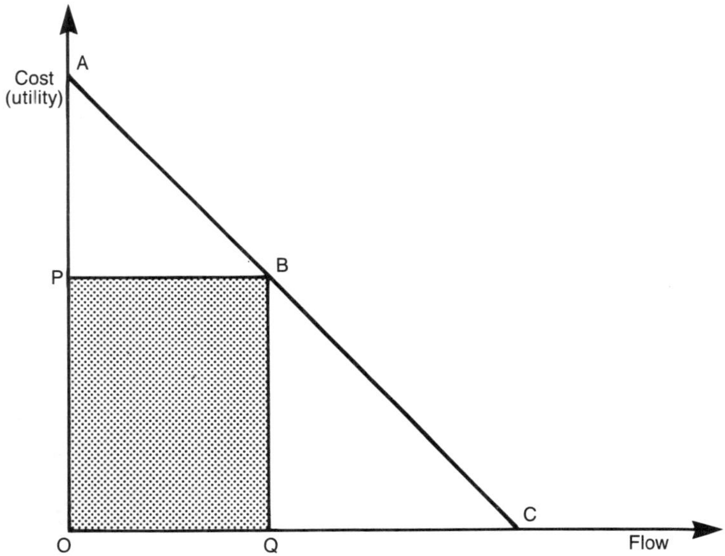

Figure 6.2 The cost/flow relationship for a footbridge

If a tax is levied at a price P, all flow to the right of the flow level Q will be suppressed, thus losing benefit given by the area QCB. The tax revenue will be pq, the area shaded in the Figure, OQBP. The travellers who would have been willing to pay more than P to travel still receive a benefit given by the area PBA.

Dupuit, as can be seen, was analysing a footbridge, which could implicitly take all the offered demand for crossings without any increase in the costs of such crossings. He was also analysing a situation where, if a trip was not made over the footbridge because of the toll, it was not made at all; there was no alternative available. The question of possible alternative destinations and of the other forms of transport available is, in cities, a key question, as I shall demonstrate later. The issue of congestion, which Goodwin and Jones call the one crucial exception, is but part of this wider question.

It is interesting that part of the reason that tolls were abolished on roads in Britain was precisely the problem spelt out by Dupuit: that tolls were often set at levels which were much higher than the cost of upkeep and interest on capital. This meant that many poor people could not afford them, and led to the Rebecca riots in South Wales in 1843.

The last Turnpike in London, following the consolidation of the Turnpike Trusts north of the Thames in the 1826 Act, proceeded north and east from central London along Albany Street (past where I now live), Parkway, Camden Road and Seven Sisters Road, but construction ceased in 1834 and it never reached its intended bridge over the Lea Valley and the main road to Epping—because the railway age was about to begin.

The inclusion of congestion in the argument begins with a footnote in Marshall's *Principles of Economics* (1890, Book V, Chapter xiii) stating that where unit costs increased as the level of production rose, marginal costs would be higher than average cost and a 'producers' surplus would be created. The result will be that resources will be diverted from an efficient use to an inefficient use.

Whilst this was not at that time applied to roads, it provided the necessary link to the welfare economics of Pigou (1920, p.192) who developed the general case for marginal cost pricing by using an example of two roads.

Pigou argued as follows (p.194):

> 'Suppose there are two roads ABD and ACD both leading from A to D. If left to itself, traffic would be so distributed that the trouble involved in driving a "representative" cart along each of the two roads would be equal. But, in some circumstances, it would be possible, by shifting a few carts from route B to route C, greatly to lessen the trouble of driving those still left on B, while only slightly increasing the trouble

> of driving along C. In these circumstances a rightly chosen measure of differential taxation against road B would create an "artificial" situation superior to the "natural" one. But the measure of differentiation must be rightly chosen.'

Pigou was arguing that a constant measure of taxation would not be appropriate if an efficient allocation of resources were required; such a measure had to take into account the difference in the marginal costs of the two products, not the average costs.

Where unit costs reduced as the level of production rose, then a 'bounty' (to use Pigou's word) would be payable if we wished to create an 'artificial' situation superior to the 'natural' one. A system of complete laissez-faire, where a uniform rate of tax or subsidy is paid, cannot do this.

Knight (1924) amplified the point about the distribution of resources between industries with increases in unit costs as the level of production rose and industries with constant unit costs as the level of production rose, using two roads as an example, in the following way (pp.584–85):

> 'In Professor Pigou's study the argument that free enterprise leads to excessive investment in industries having relatively upward-sloping cost curves is developed with the aid of a concrete example, the case of two roads.
>
> 'Suppose that between two points there are two highways, one of which is broad enough to accommodate without crowding all the traffic which may care to use it, but is poorly graded and surfaced, while the other is a much better quality road but narrow and quite limited in capacity. If a large number of trucks operate between the two termini and are free to choose either of the two routes, they will tend to distribute themselves between the roads in such proportions that the cost per unit of transportation, or effective result per unit of investment, will be the same for every truck on both routes.
>
> 'As more trucks use the narrower and better road, congestion develops, until at a certain point it becomes equally profitable to use the broader but poorer highway. The congestion and interference resulting from the addition of any particular truck to the stream of traffic on the narrow but good

road affects in the same way the cost and output of all the trucks using that road.

'It is evident that if, after equilibrium is established, a few trucks should be arbitrarily transferred to the broad road, the reduction in cost, or increase in output, to those remaining on the narrow road would be a clear gain to the traffic as a whole. The trucks so transferred would incur no loss, for any one of them on the narrow road is a marginal truck, subject to the same relation between cost and output as any truck using the broad road. Yet whenever there is a difference in the cost, to an additional truck, of using the two roads, the driver of any truck has an incentive to use the narrow road, until the advantage is reduced to zero for all the trucks. Thus, as the author contends, individual freedom results in a bad distribution of investment between industries of constant and industries of increasing cost.

'In such a case social interference seems to be clearly justified. If the government should levy a small tax on each truck using the narrow road, the tax would be considered by the trucker as an element in his cost, and would cause the number of trucks on the narrow road to be reduced to the point where the *ordinary cost, plus the tax*, became equal to the cost on the broad road, assumed to be left tax free.

'The tax could be so adjusted that the number of trucks on the narrow road would be such as to secure the maximum efficiency in the use of the two roads taken together. The revenue obtained from such a tax would be a clear gain to the society, since no individual truck would incur higher costs than if no tax had been levied.'

A diagram of costs against flow, in Figure 6.3, on the two roads may help to make this point in another way. As flow increases from zero, the narrow road gradually fills up, with costs increasing faster and faster with each additional truck, until costs on the two roads are equal. All further traffic then uses the broad road.

Beyond this demand level, therefore, the costs on the narrow road are totally dependent on the fixed costs of the broad road, and not at all on the fixed costs of the narrow road itself.

Only if demand is insufficient to fill the narrow road up to the

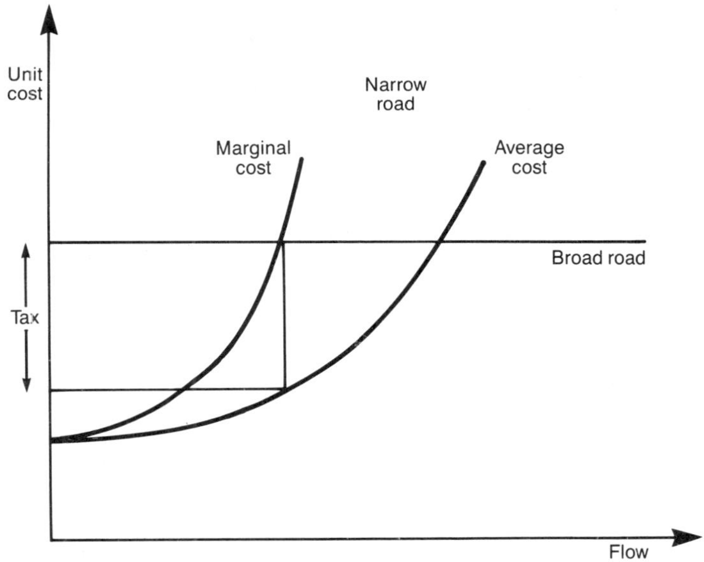

Figure 6.3 Cost/flow relationships on two competing highways

level at which some trucks use the broad road will costs on the narrow road be independent of the costs on the broad road.

The most efficient allocation of resources is where the two marginal costs are equal. To achieve this, a tax has to be levied on the trucks using the narrow road, so that, at equilibrium, the average costs on the narrow road are less than on the broad road, but the average cost plus the tax equals the marginal cost. No truck is then worse off than before, but the resources consumed are less. This is Knight's paradox.

Another relevant point to the discussion of road pricing was made by Edgeworth (1925). He argued that a tax imposed on the seller of a monopolized article may in certain cases lead to an actual lowering of the price to the buyer, although he thought it was a 'mere *curiosum*', unlikely in fact to occur and peculiar to monopoly. His example was the case of first and second-class travel on the railway, the former requiring more space than the latter and being therefore more expensive. If a tax were applied to first-class travel, Edgeworth demonstrated a case where the prices of both first and second-class travel must fall if the seller is to maximize his profit. This is Edgeworth's paradox.

Hotelling (1932) confirmed this by describing the general conditions under which this paradoxical result could be valid. Essentially, he confirmed that at least two goods or services must be involved, not necessarily from a monopoly supplier, on one of which a tax is to be levied; that the goods or services compete in both consumption and in production; that there is asymmetry of supply or of demand for the goods or services; and that there is a certain relation between the slopes of the supply and demand functions.

Essentially, the condition is simply that the product of the own-price elasticities of the two goods or services must be greater than the product of the cross-elasticities between the two goods or services, where an elasticity of demand is a measure of the rate of change of flow with respect to the rate of change of price, and is thus a measure of the slope of the demand function.

I will show later that these conditions are valid in the supply of bus and car services in a city centre. Both compete for scarce land and roads, to produce the service they offer; both compete for travellers in the prices that they offer; both have quite different costs of supply and of demand. Under certain conditions, therefore, a tax on cars the inefficient user of roadspace can produce a lowering in the price of travel for both car users and bus users.

This is precisely Edgeworth's paradox, that a tax on one commodity can lower prices for all.

Hotelling argued that:

> 'If the taxes collected from the kind of commodities we have been considering, produced with increasing cost, are used to subsidize industries operating with decreasing cost, with proper regulation to ensure that the subsidy will result in sufficiently lower prices to consumers, there may well be a net gain in benefit.'
>
> Hotelling (1932, p.609)

This is still the most concise reason for taxing cars and subsidizing buses and trains in towns, and we will spell out the evidence for this in the following chapters. The demonstration of the empirical evidence that roads exhibited the kind of upward-sloping cost curve described by Knight had to wait, in Britain, for the work by the Road Research Laboratory, spelt out by Wardrop (1952) and later by Smeed (1961). This had shown that the speed of travel on a road link fell as the volume of traffic on the link increased, thus increasing costs. And, indeed, in Wardrop (1968) the empirical evidence was extended to show that the same could be approximately true for networks. Wardrop was using Thomson's analysis of the 1966 London Traffic Survey speeds and flows in the peak and off-peak, together with his further data on Sunday traffic (Thomson, 1967a).

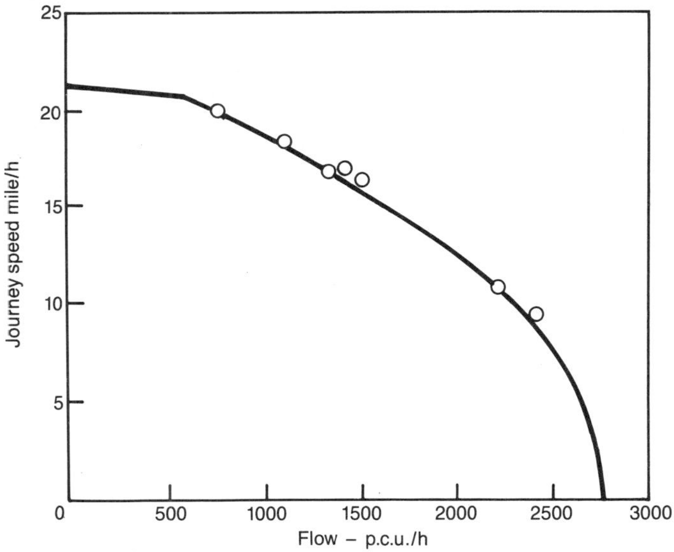

Figure 6.4 The speed/flow relationship in Central London (Wardrop, 1968)

A driver joining a traffic stream therefore incurs a cost which is only very slightly greater than the cost before he joined; he pays, in effect, the average cost. The existing vehicles in the stream are all very slightly delayed, however, by the presence of the new vehicle; the sum of these delays, the excess of marginal cost over average cost, may be considerable.

By the time that the Smeed Committee, set up in 1962, reported (1964) on *Road Pricing; the economic and technical possibilities*, there had been a burgeoning literature discussing the application of road pricing. The Committee quote a large number, starting with Buchanan (1952) and Walters (1954) and using the approach of Beesley and Roth (1962).

The Committee were, however, limited by their terms of reference to considering the pricing system relating to the use of the roads, and their attention was mostly devoted to the problem of congested areas. They specifically noted in their section on the purpose of road pricing that pricing could not cure congestion; in their view, the need for investment in urban highways would still remain (Paragraph 2.1).

Essentially, their approach used a demand and supply diagram, as in Figure 6.5. The demand curve has a decreasing flow with increasing cost, whereas the average cost curve for the supply shows an increasing cost as flow increases. The marginal cost curve thus increases at a higher rate than the average cost curve. The conditions outlined by Beckman *et al* (1956) for a traffic equilibrium are satisfied.

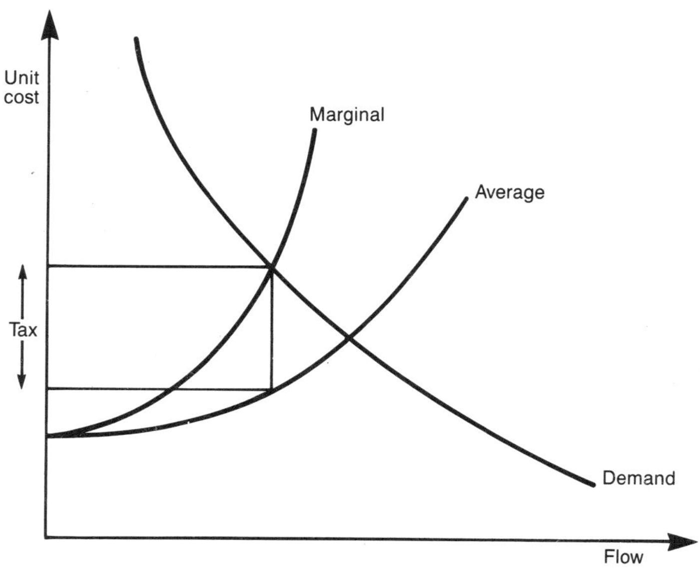

Figure 6.5 The demand-supply diagram to illustrate road pricing (Smeed, 1964)

The equilibrium point if traffic is left to find its own level is where the demand curve meets the average cost curve. If a road price is levied to make the motorist pay the marginal cost, then the demand level will fall to the point where the demand curve intersects the marginal cost curve. The correct road price to achieve this is the difference between the marginal price and the average price at the new equilibrium point.

At this point, some motorists will have chosen not to travel, thus reducing flow, and some motorists will be worse off, because they have higher costs than before. It is not surprising that such a presentation of the argument failed to convince the government at that time that road pricing was worth imposing.

Of course, the problem is that the motorists using a congested road only perceive the average cost. They are unaware that the price they have been paying for their road space has been subsidized by the fact they that have not been paying the costs they impose on all the other drivers by their presence on the road.

The road price or tax is therefore a 'congestion tax' to ensure that all drivers who do use the road pay their true costs, the marginal costs, so as to ensure an efficient allocation of resources. It is inevitable that where a facility has been provided at a lower cost than it should have been, the demand is greater than it would have been if the correct price had been applied, so that imposing the correct price reduces demand.

It is interesting that the first attempt to apply these principles was in Hong Kong, where the TRRL was involved in a transportation study. Unfortunately, the Chinese 'Cultural Revolution' of 1966 put paid to any experiments at that time though, as we shall see, the experiment resurfaced again much later, though in a different form. The analysis by the Smeed Committee implicitly assumed that all roads in the congested central area were going to be affected equally so that one could treat the network speed-flow curve as representative of all the roads in the area. It ignored the question of how people got to and from the central area.

Vickery (1967) introduced a paper on the optimization of traffic and facilities by stating:

> 'There are probably few areas in modern economic society where conditions are as far from ideal as in the congested

traffic and transportation facilities of our great modern metropolitan conurbations. This is equally true in the short run, in terms of making the best use of the facilities we have and, in the longer run, in terms of the appropriateness of the facilities for current and projected traffic needs.

'This relative inefficiency can be attributed in large measure to the fact that the individual user, faced with alternative ways of achieving his objectives, does not, under existing conditions, receive any obvious indication of the costs which his choice will impose on others, whether by impairment of the quality of service or by the cost of expanding the facilities to the point where this impairment is prevented.'

He went on to argue that, if optimum congestion tolls were charged, the motorist might well choose 'to use mass transit, to join an auto pool, to make the trip at a less congested time, or to use a less congested but perhaps more circuitous route'.

His most important point was as follows:

'To begin with it is perhaps worth observing that sometimes a facility becomes worthless precisely because it is free. For example, where a high-speed or short-cut facility of limited capacity has as an alternative a more circuitous or slower route with ample capacity, free operation may mean that a queue builds up during heavy demand periods at the access to the faster facility until the time required for queuing and transit is equal to the transit time by the circuitous route; under these circumstances no-one is able to make the trip any faster than if the faster route did not exist. Enlargement of the faster route may be a complete waste of money unless the route is enlarged sufficiently to take care of all traffic that might offer.'

This argument is precisely the point which had been made by Knight and by Downs, which was ignored by the Smeed Committee. Smeed (1961, p.24) had in fact quoted Pigou (1920) as suggesting that 'vehicles using congested roads should be taxed', but his reference to the title of the book as *Wealth and Welfare* suggests that he had derived the reference from one of his other references.

Vickery (1969) went on to discuss the proper investment strategy under congested conditions, amplifying his previous arguments. One comment is worth noting:

> 'If the use of private automobiles for access to the cores of large metropolitan areas is so attractive, under uncongested toll-free conditions, relative to other modes, that in effect nothing stops the growth of such traffic but congestion and delay, then such traffic will grow until sufficient congestion and delay are generated to constitute a deterrent until the core begins to suffer from gangrene, at which point a cumulative decline may well set in that may be difficult to reverse, even with a belated introduction of appropriate toll graduation.'

Although there is no explicit recognition that mass transit is the alternative that sets the limit for travel times, the insertion of the 'other modes' in the argument seems to imply that the congestion and delay will grow until there is an equal attraction as between road and rail or the best of the mass transit alternatives.

This paper is also noteworthy for its specification of the time-varying tolls that should be charged if it were desired to eliminate congestion at a bottle-neck during a peak period when the facility was overloaded and queues formed.

With this background, it is now time to turn to the transportation model development in London, to see how the various models were developed to incorporate the possibility of testing the effect of road pricing.

Four papers should, however, be mentioned first, as they showed quite clearly the benefits that could be achieved, particularly by bus passengers, if a reduction in car traffic could be made by some such pricing method. While still limited to road traffic, at least an alternative—bus—was being brought in to the calculations.

The paper by Smeed and Wardrop (1964) using Holroyd and Scragg's (1964) data showed that, with successive reductions in the flow of car traffic, and a switch to bus travel by the displaced drivers and passengers, it would be possible to improve the overall journey speeds of all travellers, both those left in the car and, beyond a certain reduction in traffic, even those transferring to bus. Access times by car were crucial to these calculations, and Smeed and Wardrop assumed both five and 10 minutes as alternatives.

Further papers by Sharp (1967), Goodwin (1969) and Webster and Oldfield (1972) repeated these calculations from a somewhat different angle using later data, and came to the same conclusion that overall journey speeds could be increased. Sharp used Leicester data. Goodwin extended the area covered by using Munt and Woodhall's (1967) data on journeys to the centre, not just within, and—following Smeed and Wardrop's assumption of five and 10 minutes for car access—showed that the conclusion held for these longer journeys as well. Webster and Oldfield were the first to note, using an access time for cars of 10 minutes, that car direct speeds in central London could be five miles/h (eight km/h).

None of these papers discussed the pricing mechanism by which the results could be achieved, but it is obvious that Edgeworth's paradox is relevant.

A first attempt at road pricing calculations, but a non-network one, should be mentioned as a precursor. Thomson (1967b) relied on elasticities of demand at different journey lengths for different kinds of journey to estimate the effects of two different proposals for traffic restraint in central London.

He used the London Traffic Surveys data from 1952 to 1966 (given in Table 2.3) together with the flow data collected simultaneously. The flow data was collected by counting all the vehicles passed going in the opposite direction when travelling round each circuit once in each direction.

This is shown below in Figure 6.6.

Thomson showed that the marginal cost was about three times the average cost for the central London road network. This calculation is a little spurious, since the calculations are done by comparing two flows and speeds at different times of the day, peak and off-peak, when the alternative public transport system is slightly different. We will return to this point later.

The first attempt by the GLC at a road-pricing analysis on its road network was based substantially on a schematic network model (CRISTAL) developed at the TRRL by Tanner *et al* (1973) and partly based on the further analyses by Lynam and Everall (1970 a,b) of Munt and Woodhall's (1967) work.

CRISTAL uses an idealized network for London, consisting of a set of radials and rings. Both road and rail networks are included, with bus networks on the roads. Lynam and Everall used Munt and

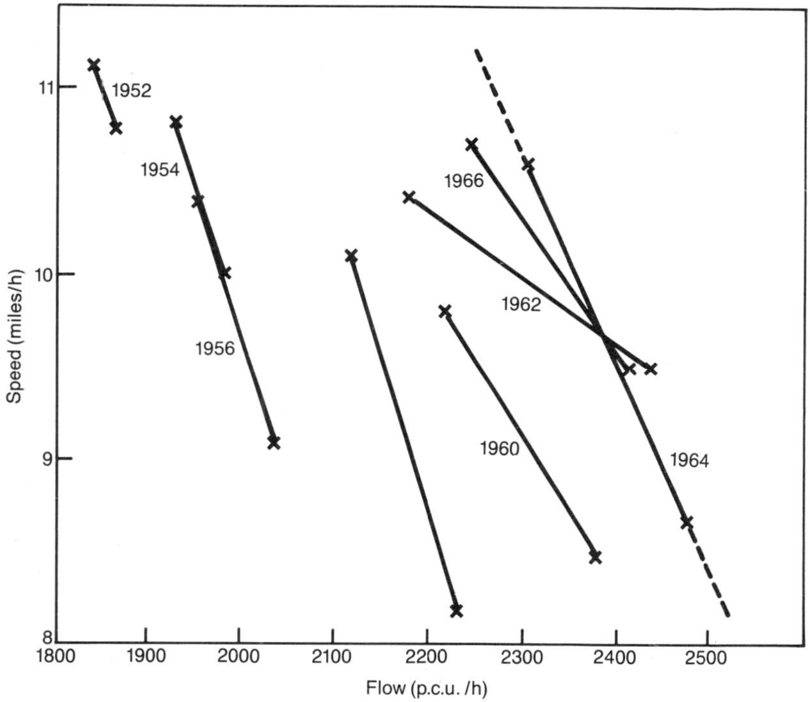

Figure 6.6 The speed/flow data from the LTS 1952–66 (Thomson, 1967b)

Woodhall's sample of 240 LTS trips, resurveyed them to separate out the walking, waiting and in-vehicle components of the trips, and compared them to scheduled service speeds and road speeds, using 40 bus route, 6 BR and 6 LT rail line schedules, plus a random set of pairs of rail stations within the Circle line. The resulting in-vehicle speeds were then used to give the required components of journey speed in the radial and ring directions for the model.

All travel in CRISTAL begins and ends at a node on the network. Standard transportation models are used for trip distribution, modal split and assignment, and for speed/flow relations on each link. (Single link models representing the speed/flow relationships of four areas were also used separately.)

The CRISTAL report makes the following comment about the difficulty of estimating the modal split parameters from the observed data:

> 'It seems likely that the inconsistency (in the estimates of the parameters) arises because the analyses based on observed modal splits do not fully take into account the extent to which relative generalised costs of different modes vary from person to person, as a result of location of trip-ends in relation to public transport services or a higher or lower than average value put on time. Thus more people may be captive to their current mode than the analysis allows for. To what extent this is also true of the CRISTAL model, in which stratification by origin and destination, by car availability and by peak and off-peak are allowed for, remains to be seen.'
>
> Tanner *et al* (1973, p.42)

The GLC adapted this model to develop tests for particular features that were felt to be important, and to test the sensitivity to charging policy (GLC, 1974).

The outputs from CRISTAL were used as an input to the road network model used in further tests. This was the Inner London Bus Priority Model (ILBPM) (Coombe, 1973) in which the road and bus networks in inner and central London were described in detail, but outer London more coarsely, and in which rail was omitted.

Travel demand without restraint was estimated from a 1981 prediction using the LTS District Test Model (DTM), developed by the LTS consultants, factored down from the 24-hour estimate to a peak-hour estimate and expanded to give the required level of detail in the finer zones as compared with the districts.

CRISTAL was then used to obtain a matrix of restraint factors by comparing the relevant flows with and without the supplementary road price, for each pair of zones in terms of their spatial position.

These two sets of flows were then loaded on to the network incrementally, in stages of 30 percent, 30 percent, 20 percent and 20 percent respectively, re-assessing speeds after each increment. Such incremental loading allows for potentially different routes for a given inter-zonal flow.

This procedure produced problems. Despite the fact that overall travel demand was some 25 percent higher in the 1981 prediction than then existing levels, the assignment procedure with unrestrained flows produced traffic levels in the centre equal to existing ones after only assigning the first two increments; that is, 60 percent of traffic.

This level was therefore taken to represent the existing situation and the 80 percent loading a future date with a one third increase in traffic in the centre (GLC, 1974, paragraph 4.2.22).

Thus, whereas the LTS consultants had used a linear programming procedure to reduce flows to the capacity of the network, the Supplementary Licensing Study merely did not assign all the flows to the network!

Whilst a number of sensitivity tests were performed on the various parameters used in CRISTAL, one comment by the GLC study is particularly worth noting, namely:

> 'CRISTAL uses a linear speed/flow relationship, which tends to underestimate speed increases for low levels of restraint, and overestimates them at higher levels'.
>
> GLC (1974, paragraph 4.3.17)

This is what one would expect from comparing the speed/flow relation in Figure 6.4 to a straight line relation.

The analysis in this study clearly showed that a road pricing strategy was beneficial. The modelling strategy for this study, however, contained a number of unresolved issues and inconsistencies. Although this was the first time that rail as well as bus had been incorporated in the analysis, the modal split models were very poorly specified. The continuation of the modelling work during the rest of the decade attempted to resolve some of these problems. CRISTAL was, however, not developed any further, but was abandoned by the TRRL, although as a non-network model it offered a rapid tool for analysis and thus gave some insights that could not be gained with the full complex procedure using a network model.

The political problem remained, however, essentially the same as it was when the Smeed Committee had reported 10 years earlier. The Conservative administration which began in 1969 and which had initiated the study had been rejected by the electorate in 1973 as already noted, but not before an incident had occurred which sheds some light on the pressures on them from the national government.

On a wet winter Wednesday evening in the dark in 1973, a monumental traffic jam had delayed the then Prime Minister, Edward Heath, on his few yards' journey by car from Downing

Street to the House of Commons. The then GLC leader, Desmond Plummer, was in Tokyo, attending a conference on the future of the major world cities. A message was sent, demanding that something must be done about London traffic!

The Labour administration of the GLC returned in the 1973 election finally rejected the road pricing proposals in 1975 on the main basis that the imposition of a road tax was unfair to those poor people who used cars in the centre, who would be forced to pay a higher price than before or not travel.

This was despite the fact that it could be clearly shown from the analysis that the proportion of cars driven by poor people into and through the centre was a small proportion of the total of cars in the centre. Drivers in richer households were estimated to be four times more likely to use their cars in the central area (though these data were based on the 1962 LTS survey as the 1971 was not yet ready), and of course were far more likely to own cars anyway.

Moreover, it was despite the fact that, unlike the Smeed Committee, the GLC Supplementary Licensing Study showed that public transport—particularly buses—would benefit from a reduction in road traffic, confirming the results of Smeed and Wardrop (1964), Sharp (1967), Goodwin (1969) and Webster and Oldfield (1972); and that, overall, the poorer section of the community, who used the buses more, would be better off.

They also felt that the proposals were impractical in their current form, especially as no differentiation was made between 'essential' and 'unnecessary' traffic.

However, the Labour administration decided to continue its investigations to see if the objections could be overcome. A target of a one-third reduction in car traffic in central London was set as this appeared to give the maximum benefit. The political remit for the continuation of the study was thus to find a method of licensing 'essential' traffic, who would thus avoid the tax, whilst obtaining the necessary reduction in traffic flows.

It is worthwhile noting that the study was used as the inspiration for the Singapore Supplementary Licensing Scheme, which was put into operation in 1975, and was the first such scheme to be so implemented, and until recently, the only such.

When I visited Singapore in 1977 and gave a lecture to the University of Singapore Department of Civil Engineering Transport

Course, there were a lot of questions as to why we had not proceeded with the scheme in London, since it was so obviously already working well in Singapore!

I found it difficult to convince the students that a small section of the community who loses out on a particular change of public policy can sometimes have a disproportionate effect on the decision in a city like London.

In parallel with the Supplementary Licensing Study, the London Rail Study had been set up by the Minister for Transport Industries of the national government together with the GLC in February 1973. The report of the study is known as the Barran report (Barran, 1974), after its chairman.

Their brief was to review the arrangements for passenger rail travel in London and the commuter region. They decided to examine three hypotheses about future trends in population, employment and travel:

> 'Hypothesis A—continuation of present trends.
> i) low Greater London population
> ii) existing predominantly centralised employment pattern
> iii) declining level of public transport use because of declining levels of service, and growth in car ownership and use.
>
> 'Hypothesis B—median projections
> i) median Greater London population level
> ii) median employment patterns
> iii) increasing use of public transport because of improved levels of service, and severe restraint on the use of the car.
>
> 'Hypothesis C—change of present trends
> i) high Greater London population
> ii) decentralised employment pattern
> iii) increasing use of public transport because of improved levels of service, and severe restraint on the use of the car.'
>
> Barran (1974, paragraph 11.5.5)

The latter hypothesis reflected the GLC policy statement of October 1973—'London: the future and you' (GLC, 1973).

The method of analysis was based on the 1966 journey to work data from the 1966 census, a 10 percent sample comprising over

400,000 trips, defined by the main mode of the normal journey. These were factored to give actual peak-hour trips. The 1971 GLTS data was not ready by the time of the analysis.

Initially, a forecast of population and employment changes up to 1991 was prepared, and the travel patterns held constant; that is, the same proportion of public transport trips between areas was maintained as in 1966. The subsequent forecasts of travel were made using the redistribution and modal split models of the LTS consultants.

The report continues:

> 'This tended to increase passenger traffic by something under 10 percent in the case of new schemes in areas which already had rail services, and by 15 to 20 percent in areas which had none previously. This appeared to accord with common sense, and with experience on the Victoria Line (opened in 1969–71).'
>
> Barran (1974, paragraph 12.5.3)

Both forecasts were assigned to the various test networks using the TRANSITNET assignment procedure.

The initial forecasts using the median hypothesis suggested declines of 16 percent by 1981 and 24 percent by 1991 of morning peak-hour rail journeys to central London, compared to 1966, with the Underground declining more steeply (28 percent and 37 percent) than the BR services (six percent and 13 percent). Hypotheses A and C suggested declines of 19 percent and 27 percent in the same journeys by 1991 (Barran, 1974, paragraphs 12.6.3–7).

With these sorts of decline envisaged, it was no wonder that the proposals considered to improve services were in general found to be not worth the investment. A re-run of the LTS proposals, including the Cross-Rail scheme, was done, together with a large number of others culled from over 100 proposed.

Three increasingly expensive investment packages were discussed. Most expenditure was concerned with replacement or renewal of stock and lines, but a number of proposed new lines were approved. Extension to the Fleet/River Line (now the Jubilee Line) from Strand (now Charing Cross) via Fenchurch Street and Docklands to Woolwich and Thamesmead was approved in the basic package. The next had a scheme of through-running on British Rail lines between north and south London (the Thameslink service) and the Chelsea-Hackney Underground line. The final package had an extension of the Fleet Line via Lewisham to Hayes and Addiscombe after 1990, if it were still justified by traffic demand.

The study recommended further work on the Cross-Rail project and that its alignment should be safeguarded. As we shall see, this eventually became the Central London Rail Study of 1988.

They also recommended further work on a zonal fares system and on automatic fare collection, both of which had to wait for the Labour administration of 1981 before implementation.

The major problem with this London Rail Study was that the forecasts of employment change were wrong, and indeed, after 1982, in the wrong direction. This has made a substantial change to the rail system necessary as we shall see.

It was, however, as a result of the Barran report that I began in 1974 to think that our methods of estimating future demand had some fundamental flaws in them, if they could not justify the worth of the Cross-Rail scheme. In this, I was influenced by the way that Paris had decided, in 1965, to go ahead with a Regional Express Railway through the centre of Paris, much like the Cross-Rail project, and the considerations brought to bear on that decision. That, however, must wait till the next chapter.

The main strategy for the second attempt at road pricing analysis was the use of the STEM suite of programmes (GLC, 1983). These were developed from the DTM, the road model, and COMPAS (COMPuter ASsignment), the public transport model, to enable analysis of the 1971 survey to be done. COMPAS was a GLC development of TRAMP, a public transport model developed by consultants, which had essentially the same features as TRANSITNET.

The zonal boundaries were quite different to those of the LTS, given the political reorganization of the area in 1965 and the formation of the GLC, although approximately two-thirds were unchanged.

The network was also substantially different, given the different priorities of the GLC in its road strategy from those of the Department of Transport which essentially defined the earlier road network. The road network was, however, still defined at district level (like the DTM), whereas the public transport network was defined at zonal level like TRANSITNET.

The survey was also a uniform household survey, apart from higher sampling rates in some areas, whereas the 1962 household survey was, as already noted, a non-uniform survey taken from two sampling frames.

Unlike the 1968 LTS, the road network was assessed on a peak-hour basis, just like the public transport network in the LTS, but the sets of trip ends were derived directly rather than being factored from the 24-hour flows as in the LTS.

Although calibration of the STEM model could therefore be done on the 1971 network and household data, it was not possible to use the 1962 data set for verification, since in effect the 1971 model was begun from scratch. In fact, the calibration was finally done on 1976 network data, since the model was not fully operational until then.

There are two main features of this model worth noting. The first is that distribution and modal split are performed concurrently, not sequentially, and that there are a number of deterrent functions for areas and trip purposes, in particular for central area work trip-ends as opposed to non-central. These deterrent functions have the effect of reducing the number of trips to trip-ends which are more costly to reach, as required by Beckman *et al*'s criterion.

The fact that they are, however, applied at the origin end rather than the destination end means that some central work trip-ends are more attractive than non-central work trip-ends at the same cost of travel, which, as already noted above, is not strictly permissible if a unique equilibrium is to be achieved in the travel pattern. The details are given in Hathaway (1975), but are not noted in the final report of the study (GLC, 1983).

The integration of distribution and modal split, although with different deterrent functions, avoids the drawback of the earlier models: that the road trips, once fixed, did not then interact with changing costs on the public transport system and vice versa. It does, however, have other drawbacks, which are discussed by Senior and Williams (1977), essentially to do with the perception of composite costs. We will return to this topic later.

The second is that a linear programming procedure is operated on central area parking costs to force the number of car trip-ends in the centre to equal the number observed in 1971 (thus adding notional costs to observed costs). This point is described in GLC (1983, paragraph 5.2.7), and is a more sophisticated way of dealing with the same problem which was encountered by the LTS consultants in the 1962 data. These notional costs were not analysed to see if they bore any resemblance to access times which we have discussed earlier; they include both time and money; that is, parking costs.

As in CRISTAL, the speed/flow relationships used in the STEM model were very primitive, being no more than a set of straight lines for each class of road. A maximum speed was defined at zero flow. This remained unaltered as flow increased up to a certain limiting value, after which it declined linearly with increasing flow to a minimum value, and then it remained constant however much flow increased.

This relationship was used because of the computational limitations; any more complicated relationship was much more expensive to use and it was thought in 1971 that it would not add much to the accuracy of the results. It is still the recommended relationship in the Department of Transport's *Traffic Appraisal Manual* (1981).

A cost could also be added in to each link, representing the cost of driving along the link at that speed, and converted via a value of money in time units to a generalized time.

With these speed/flow relationships, one could then repeat the computer procedure by assuming a speed on the roads, assigning the traffic to minimum paths, finding the flow assigned, reassessing the speed on the roads, and repeating until a stable pattern was formed.

This procedure was called 'capacity restraint'. With an all-or-nothing assignment the procedure tended to be unstable (i.e. oscillated between high and low speeds on particular roads in successive iterations), but with a judicious choice of the proportion of traffic assigned each time, the equilibrium position could be approached reasonably quickly.

Thus the first criterion laid down by Beckman *et al* (1956) for an equilibrium travel pattern did in fact seem to be satisfied by a road transport network for car and freight travel with such speed/flow relationships.

Initially on the public transport assignment, whilst the speed on the links and routes did not depend on the flow, fares were roughly estimated by developing a relationship with route distance, rather than taking the cost from individual zone-to-zone assessments, which would have involved a tremendous amount of work. A value for money in time units was applied to fares which was specific to public transport passengers, although there was an argument that suggested that in the evaluation process the value should be the same as for car users for equity reasons. This procedure also satisfied Beckman *et al*'s criterion.

The STEM, although a network model including the rail network, used only one public transport assignment, incorporating both bus and rail together. Public transport modal split was actually addressed with the SCENARIO model (Fairhurst and Smith, 1976), since the public transport assignment model was poorly calibrated (GLC, 1983, paragraph 5.2.20).

The STEM gave the wrong response in modal split between bus and rail for a fare change on one of them. In particular, the bus link speed in the public transport assignment model was independent of the link speed in the road assignment model. This was a crucial drawback in a model being used to test the effect of road pricing on road speeds. SCENARIO had, therefore, to be used to estimate demand changes (paragraphs 5.1.17 and 5.1.18).

The SCENARIO model is, however, a non-network model with particular emphasis on market-share, or modal split. It was developed by London Transport to enable them to test strategies for their bus and rail systems. The model splits the GLTS area into market areas, one such being trip-ends in the central area.

The same difficulty in assessing modal changes was found in the SCENARIO model as in the CRISTAL model. This was countered by the deliberate introduction of a captive/non-captive element into the process of estimation to damp down the switch of passengers. The calibration of the modal split parameters relies essentially on the short-term reaction of travellers to fare or price changes and to changes in the level of service; especially on the work of Lewis (1977). There is essentially no redistribution element, though some element of redistribution may well be incorporated in the calibration of the modal split parameters (GLC, 1983, paragraph 5.2.25).

Lewis's work is particularly interesting here, in view of Hotelling's (1932) demonstration of the conditions for Edgeworth's paradox to hold; that is, that the product of the own-price elasticities should be greater than the product of the cross-elasticities. Lewis analysed total vehicle flows into central London across 10 automatic traffic counters, over the four-year period 1972–75 (i.e. including the first oil price crisis), and attempted to correlate the variations in flow not only with fuel price (which varied considerably over the period) but also with public transport fares and vehicle-kms operated, both buses and LT rail. The data were disaggregated into weekday peak and all-day, and Saturday and Sunday.

The following is quoted from the discussion of Lewis's work in TRRL (1980, p.121).

The elasticities of traffic flow with respect to fuel price in the different periods were all highly significant, and that in the peak (–0.04±0.02) was only half the overall Monday to Friday value (0.09±0.02) while the weekend elasticity was much larger still (–0.35±0.10). The elasticity of bus passenger-km with respect to bus fares was not given in Lewis's work, but is taken from TRRL (1980, p.119) using LT data (Rendle *et al*, 1978). The peak fares elasticity (–0.27±0.04) was slightly smaller than the Monday to Friday value (–0.32±0.02), assuming both bus and rail fares are increased in step. (There is a further transfer component [–0.11±0.03] to rail if rail fares are kept constant.)

The cross-elasticities with respect to price were higher in the peak (0.06±0.03) than in the overall Monday to Friday value (0.03±0.03), suggesting that competition is higher in the peak when public transport services are better.

Thus the Hotelling condition for the Edgeworth paradox was met in Lewis' peak data (–0.04×–0.27 > 0.06×0.06). This was not noticed at the time or indeed later until now.

The major problem with calibrating the road assignment in the STEM arose from the access costs. As noted in earlier chapters and especially Table 4.22, access times for cars in the central area appear to be much higher than assumed in the STEM. A parking sub-model was therefore derived which produced notional average costs of parking and access (STEM operates in generalized time) to ensure that the demand observed was matched. It was recognized that:

> 'A number of distinct groups of car users exist where parking usage is considered'
>
> GLC (1983, paragraph 5.2.22)

though nothing was actually done to represent these different groups, paying different costs.

A doubly constrained distribution/modal split model of the STEM type, where trip origins and destinations are both fixed, where costs are increased on one mode, will show a shift of flow from the area where costs are increased on that mode to other areas, and its replacement by other flows on the competing mode.

It cannot show, for flows are not separated in that way, a shift in the components of flows such that those individuals who value time highly are hardly affected, whilst those who value money highly are strongly affected. The former are, in general, the richer members of the community and the latter the poorer.

In principle, however, a model such as the STEM could be further disaggregated to include such effects, proxied perhaps by an income categorization as we have just suggested. The STEM network model was therefore a considerable advance on the previous models used, and was far more consistent and thorough in its analysis, although a number of problems still remained to be resolved.

Although the second set of analyses were finished in 1979, and a report prepared under the title *Area Control*, this was after a new Conservative administration had taken over at the GLC in 1977. There was some trouble over publication of the results. Bayliss (1979) and Prestwood Smith (1979) gave presentations of the results to professional and international audiences, and Traffic Engineering and Control (TEC) (1979) published a summary, but the administration refused to publish the full report.

Horace Cutler, the Conservative leader, decried the very idea of road pricing. He argued, apparently before examining the study's report, that there was no evidence that the idea of imposing what, in effect, amounted to an additional tax on motorists would actually be workable. He went on, according to TEC:

> 'Accordingly, the idea has been dismissed as both unfair and impracticable, and will not be considered again by the present administration other than in the most exceptional circumstances'.
>
> TEC (1979, p.366)

These two emotive descriptors—unfair and impracticable—had actually been precisely the criteria which had caused the previous

Labour administration to continue with the study of road pricing to see if they could be overcome.

The report had therefore addressed these issues rather carefully, and concluded that a licensing scheme could be prepared to achieve a one-third reduction of central area traffic which was indeed practical and that, far from being unfair to poorer people, they were on the whole likely to benefit from the increased bus running speeds, as the earlier report had suggested.

The report also showed, on the basis of the STEM model calculations, that, for a given price, approximately equal numbers of former central London car travellers would change mode or change destination outside the central area. It was assumed that destinations so vacated (i.e. mostly workplaces) would be taken up by other public transport users. This is an inevitable consequence of a model with fixed trip ends.

The main Conservative opposition seems to have been on the basis that the imposition of any charge would adversely affect small businesses in the centre and lead to a further decline in business activity. This was at the time of great and rising unemployment, when manufacturing industry in central and inner London had been declining at a great rate for some years. A full analysis of the decline in employment in central London is available in Evans and Crampton (1989). This shows that the peak of employment in central London was in 1962, from whence it fell for 20 years to a low in 1982. Since then, it has been rising rapidly again, but not in manufacturing or indeed in wholesale trades. The great rise has been in information and money processing industries.

The report was finally formally published in 1983 after the incoming Labour administration of 1981 had decided to go out to further consultation on the issues raised.

The GLC itself was, however, abolished by the national government in 1986 before the results of that consultation could be assessed, and the work taken further.

It has been left to the London Planning Advisory Committee (LPAC) to carry the work on, and I will discuss their contribution in Chapter Eight.

It is worth noting, however, that this, the second London study of road pricing to show its value, was taken up by the national government and applied to Hong Kong. An electronic road pricing system was devised and tested (Dawson 1983) and demonstrated to be practical. Electronic 'number plates' were read as cars passed over detectors in the road at the boundaries of cells in the network, and recorded on a central computer. The car owner would then be billed once a month, just like a metered telephone or electricity user.

The scheme was, however, put into abeyance following opposition from those who felt a central computer able to track individual car movements would be an invasion of privacy, a concern given even more force now with the agreement to return Hong Kong to China. We will note how this privacy problem can be overcome in Chapter Eight.

But now a new factor entered the analysis, or rather an old factor made itself known through the kinds of analysis outlined in Chapters Two-Four, and to this we now turn.

7 *The Downs-Thomson Paradox*

Pigou (1920) argues that for maximum efficiency of the use of resources a tax has to be applied to the goods or services produced with an upwardly-sloping cost curve ensuring that the marginal cost of production equals the average cost of production plus the tax. He applied this to two roads as an example, and we have seen how the development of the case for road pricing as a 'congestion' tax has developed from there.

Knight (1924) extended the case to two roads, one of which had an effectively zero slope of the cost curve, so that, beyond a certain volume of traffic, the cost of using the road with lower fixed costs did not depend on the flow on that road but on the higher fixed costs of the road with zero slope of the cost curve. This point was amplified by Vickery (1969) as part of his argument against building freeways in congested conditions. Knight's paradox is that levying a tax on the road with an upward-sloping cost curve reduces the resources consumed.

We have seen how the GLC models developed in the 1970s to incorporate into the road pricing argument both redistribution between destinations and modal split between alternative methods of travel. The crucial element missing from these studies is the reaction of the public transport operators to a change in the demand level as travellers change from car to public transport. Effectively what the models have assumed is that the cost curve for public transport has a zero slope, just like Knight's broad road.

We now move on to consider the case of goods produced with a downward-sloping cost curve. Pigou argues that for maximum efficiency of the use of resources these goods should have a 'bounty' or subsidy applied to them to ensure that the marginal cost of production equals the average cost of production minus the subsidy.

Edgeworth's (1925) paradox, as developed by Hotelling (1932), was that for two competing users of space—one with an upward-sloping cost curve and the other with a downward—a tax applied to the former or a 'bounty' to the latter could reduce the price of both by reducing the resources consumed.

Lardner (1850) was the first to recognize that railways have such a downward-sloping cost curve; or, putting it a different way, there are economies of scale in railway operation. He recognized the problem of allocating track costs to the different kinds of traffic upon the railway (1850, pp.38, 52). Joy (1973, p.100) comments:

> 'The advantages of the potentially high capacity of railway route are accompanied by the disadvantages that the costs of providing the route are related primarily to its capacity, rather than to its utilisation. This creates two related problems. First, the over-provision of route capacity, relative to traffic requirements, will lead to track costs which are too high to be recovered fully from users. Secondly, once track is installed the invariability of its costs to traffic volumes makes difficult the attribution of track cost to a particular traffic. The allocation of track costs to traffics is perhaps the oldest problem in the field of railway economics.'

In Joy (1989, p.47), an invited paper in the issue of the *Journal of Transport Economics and Policy* commemorating its 21st anniversary—for which I was the guest editor—he has taken the argument further:

> 'Railway policy making was profoundly affected by the belief that a very high proportion of railway costs were fixed, and that, as changes in volume would affect only the small proportion of costs which varied with output, rates for new traffic need only cover those obviously variable costs (Acworth and Stephenson, 1924; Ripley, 1916; Wellington, 1889). This argument depended on three conditions:
>
> (1) that the costs of a railway *did* include a high proportion of fixed costs;
>
> (2) that only *new* traffic was priced to cover only the "low" variable costs;

> (3) that the revenues of all the other traffics covered *all* the other costs.
>
> If the third condition was not met, the railway would be unprofitable. If the second was not met, gradually all existing traffic would threaten to leave unless it too was classed as new traffic. If successful, this threat would then violate the third condition. But the most important condition is the first: the whole argument depended on the validity of the assertion that most railway costs would remain unchanged in the face of changes in traffic volumes. This type of argument has been used in different ways to influence railway policy in different countries.'

Thus Joy is arguing that the real cost to the railway of carrying extra traffic in the use of extra resources is the marginal cost of the extra flow or volume. If, however, the railway attempts to charge only this amount, and not cover its average costs, it will eventually go bankrupt. It follows that a subsidy from the government, on behalf of society, to cover the difference between the marginal and the average costs will lead to the most efficent use of resources.

Turning now to urban passenger-carrying railways, which do not have freight or long-distance rail passenger movements to complicate the apportionment of fixed costs, Lang and Soberman (1966) give an analysis of urban rail transit systems in the USA. This analysis suggests that there are indeed economies of scale in such operations, and therefore that marginal costs are below average costs.

Even in such systems, there are difficulties in apportioning costs, particularly to the peak and off-peak services, owing to: the different lifetimes of the various track and terminal elements, and the power and communication systems; and to the different maintenance and operation costs of the different services.

It is a lot easier to apportion costs in a bus system since there is none of the complication due to the operation of the exclusive right-of-way. Mohring (1972) gave a detailed analysis of the average and marginal costs of operating the Minneapolis-St Pauls (Twin Cities) bus system. He also used some data from London (on passenger boarding times), supplied by Coburn of the RRL.

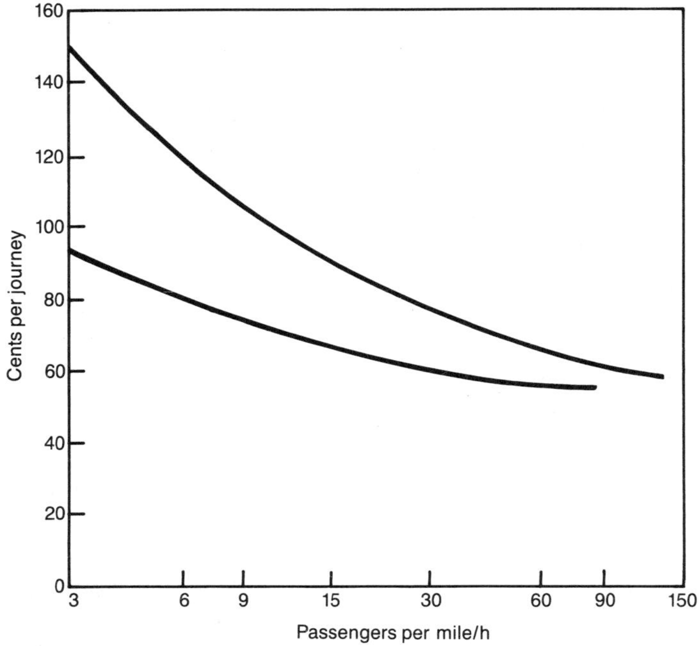

Figure 7.1 Average and marginal cost/flow relationships (Mohring, 1972)

His Figure 2 is partially reproduced here as Figure 7.1. This shows Mohring's calculations of the average and marginal costs for a steady state route (equal flows in both directions) in the peak-hour. He also studied the main and back-haul directions of tidal flow routes (with flows in the ratio of 5 to 1), and of off-peak routes.

Notice that flow is plotted logarithmically, not arithmetically. At zero flow, average and marginal costs are equal. At high flows of passengers, beyond the range studied here, the congestion effect of buses impeding buses becomes important, as we shall see later. Notice also that the flow rate is expressed as passengers per mile-hour. Mohring assumed passengers were uniformly distributed along a bus route, and that the average trip length was three miles. He also assumed that walking and waiting times were valued at three times in-bus time, following Beesley (1965). Mohring also assumed a bus speed of 20 miles/h.

Mohring concluded that, for the particular operation he studied:

'Going from current to optimal operating characteristics would call for the following:

(a) reducing peak period bus headways about a third
(b) reducing off-peak headways by about a half
(c) reducing the current fare by about 40% during the peak
(d) and about 75% during the off-peak.

Putting these changes into effect would require increasing the present bus fleet by about 42% and operating about 70% of these buses during the off-peak period rather than about 35% as is presently the case.

'If no increase in patronage were to result from these changes, the bus system would generate revenues sufficient to cover about 40% of its total costs.

'However, these changes would reduce trip prices considerably, by about 20% and 30% respectively in the peak period main and back-haul directions, and by 35–45% during the off-peak period. Patronage would almost certainly increase thereby making further service increases desirable and

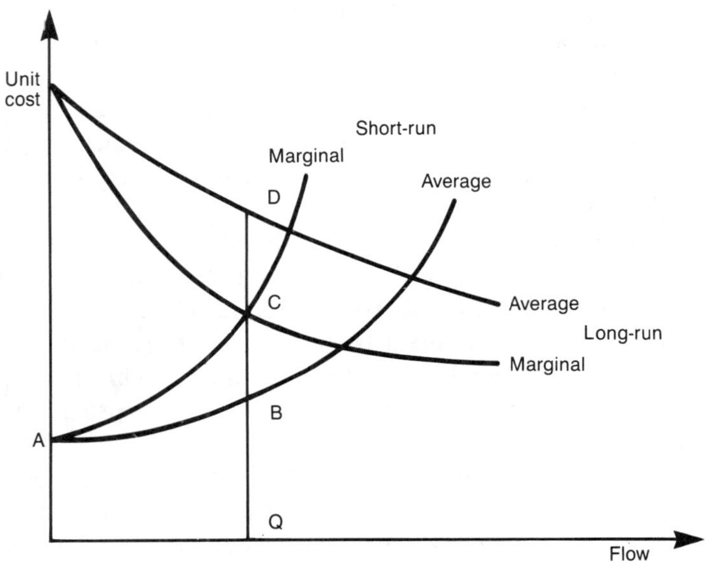

Figure 7.2 Average and marginal cost/flow relationships (Mohring, 1972)

> reducing the system deficit somewhat, at least on a per passenger basis.'

Thus we can see that Mohring clearly argued the case for subsidy, to ensure that the passenger paid the marginal costs of operating a bus system, whereas the operator received the average costs, in order to ensure that the system operated at the most efficient level in its use of resources for the given demand.

We can take Mohring's argument and apply it to the London data we gave in Chapters Two-Four. Figure 7.2 is a partial reproduction of Mohring's Figure 1, which includes the short-run average and marginal costs as well as the long-run average and marginal costs we have just been examining. At a given demand level, the most efficient use of resources is at the point where the long-run marginal costs equal the short-run marginal costs: the point C. This is the price at which the goods or services should be sold.

It will be seen that the short-run average and marginal costs have precisely the same form as the cost curves for car travel discussed in the previous chapter. This is because with road pricing we are essentially dealing with a fixed road system and attempting to allocate resources efficiently on it. The long-run costs become important only when we are trying to estimate the size that the road system should be.

In public transport on the other hand, we are attempting to estimate the size of the network and the fleet of vehicles that we need to service a given demand efficiently; so we need to examine the long-run costs.

In the short run, as flow increases (but the network and the fleet of vehicles operated stays the same) the level of service will deteriorate. Mohring differentiates two factors. The extra passengers take longer to board and alight from a service, thus increasing the running times of the service. The fixed number of vehicles then are able only to operate at longer intervals—thus increasing waiting times of passengers.

This situation will be recognised as part of the current problem in London. The number of passengers to central London on the railways, but not the buses, has increased markedly since 1982 without a corresponding increase in the number of trains, and the railway level of service in the peak has deteriorated as a consequence.

The number of passengers on the buses, on the other hand, has in general continued the long-term downward trend (Evans and Crampton, 1989). The situation on a number of the bus routes has been complicated by the removal (as part of a cost-cutting operation) of the conductor who took the fares . When the driver has to take the fares, boarding times are increased. London Buses have had to reschedule their central area bus routes with driver-only buses to add a further 20 percent extra time to the journey. This increase in journey time has undoubtedly been a factor in reducing demand.

Mohring goes on to interpret Figure 7.2 in the following manner. For a given demand level OQ, the cost per passenger of the in-bus journey is given by the intersection of the long and short-run marginal cost curves QC. This cost comprises two components.

The first—the average short-run cost, QB—is the cost per passenger of the in-bus journey in money and time of all the passengers being carried in the bus, plus the cost of picking up and putting down a typical passenger. The time cost of such a journey equals the time cost of an equivalent car journey.

The second—the 'congestion' cost, BC—is the cost imposed on each passenger by the fact that the bus has to keep stopping to pick up and put down other passengers. This increases the travel time of all passengers, and reduces the distance that can be run with a given bus, thus increasing the cost.

The difference between the long-run marginal and average costs, CD, is therefore the time cost of the passenger who has to walk to and from, and wait for, the service or services on his or her journey.

Some idea of the order of magnitude of these various costs for an average journey of 2.7 km (1.7 miles) in the 10-km square covering central London can be seen by just taking the time costs as given in Chapter Four using the component equations of Table 3.3. The in-bus time is 14.2 minutes, the walking time is 12.2 minutes and the waiting time is 6.4 minutes, giving a total of 32.8 minutes. The in-car journey of this length at 9.2 miles/h (Table 2.1) would take 11.1 minutes, or 78 percent of the in-bus journey time.

QB is thus 11.1 minutes, BC 3.1 minutes and CD 18.6 minutes. If CD were to be the subsidy we should apply to bring the long-run average costs down to the long-run marginal costs in time alone, the subsidy would be 57 percent. Because we have ignored the money

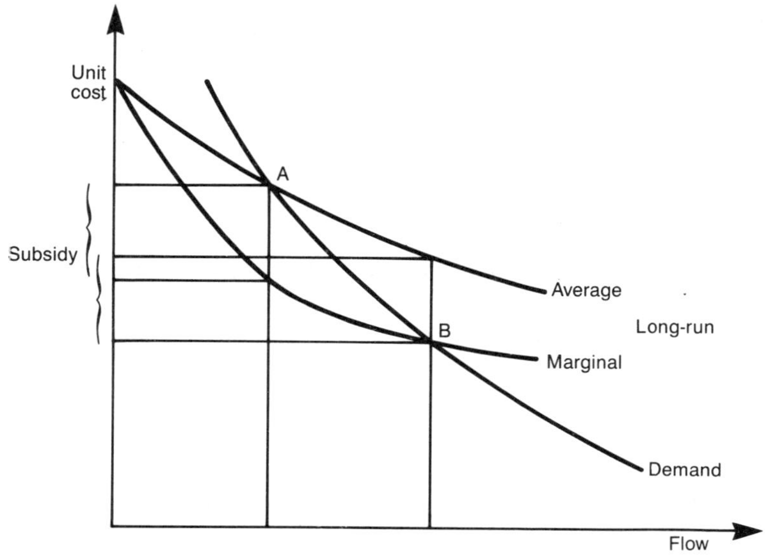

Figure 7.3 Demand/supply relation for bus travel

costs, the subsidy will be a lower proportion. However, it is generally reckoned (TRRL, 1980) that the money cost of a bus journey is about one third of the total cost, so the subsidy would then be 38 percent of the total or about 114 percent of the money cost.

This is not the end of the story. As Mohring recognised, if we subsidize bus journeys the demand level changes and more travel is made. We can see this by examining Figure 7.3.

If the bus operator has to cover all his costs, the demand level will be at the point A. With a subsidy to ensure that marginal costs are covered, demand will rise to the point B. With the higher provision of bus services, there will be a reduction in the time costs of walking to and from and waiting for bus services; so the proportion of subsidy per passenger may well be less at B than it would have been if demand were fixed at the level at A.

This proportion will also depend on the road speeds, which will be affected by any road pricing policy as we have already noted. If road speeds are increased by a road pricing policy then buses will also be able to travel faster, thus reducing their costs and so affecting the proportion of time spent walking and waiting to total costs.

Thus we can see that public transport services, both rail and bus, are services produced with a downward-sloping cost curve. As the flow of passengers on public transport increases, a public transport operator can supply more services with a lower cost per unit of output; that is, he gains economies of scale. The passengers also gain further because the increase in service frequency and reduction in network spacing reduces their access and waiting time.

Downs (1962) seems to have been the first to argue through the consequences for the allocation of the space of the city between car users and public transport users. He argues, in the terms we have just been discussing, that where roads had an upward-sloping cost curve, and public transport had a downward-sloping cost curve, an equilibrium between car and public transport costs would be established. At this point, there would be travellers who would have the same costs by road as by public transport, and who would be indifferent as to which of the two methods of travel they should use.

The cost of travel he referred to as the 'undesirability'; that is, the cost should include all the time, money and effort expended on travel which was not an end in itself but only a means of reaching a goal, the activity it was desired to undertake at the destination. The costs referred to are the perceived costs, and depend on the tax or subsidy actually paid to each mode of transport.

The travellers at the equilibrium point he called the marginal travellers. Downs, however, was speculating about a system (i.e. a network of roads and public transport), so, at equilibrium, there would still be travellers who would have lower costs by road than by public transport and vice versa depending on their precise journeys.

Downs (1962, p.405) went on to speculate on what must happen if there is such an equilibrium between public and private transport costs with a number of examples of different public transport service types. If road capacity is increased, by the construction of a radial expressway, when the alternative is a segregated-track rail service; then he argued as follows:

> 'If a very large number of persons shift from segregated track transit to automobiles, the cost of such transit per passenger

> may rise so that its attractiveness is drastically reduced. In such a case, congestion on the highways may have to become slightly worse than it was before the expressway was opened before automobile travel becomes just as undesirable as segregated track transit travel.
>
> 'But only the marginal equality of these two undesirabilities can establish a new equilibrium.
>
> 'We thus arrive at the paradoxical conclusion that the opening of an expressway could conceivably cause traffic congestion to become worse instead of better, and automobile commuting times to rise instead of fall!'

We can see that Downs recognised that it is the marginal travellers who change—the equilibrium is made up of many people for whom car is better, many for whom rail is better, and a number for whom both are equally good. For an equilibrium to be established, travellers must be prepared to change to the mode which is more attractive for themselves.

Downs called the people searching for better routes 'explorers' and the non-changers 'sheep'. I will present later some evidence from London to show that indeed there are people who change their method of travel on particular journeys within a week, as it is absolutely vital to the argument that there are such people who are prepared to change.

Although Downs published in *Traffic Quarterly*, this work seems to have had absolutely no impact upon the transportation planning profession for the next two decades, despite his drawing upon examples from both Chicago and New York (among other cities) to substantiate his argument. In part, this may well have been due to the nature of his title 'The Law of Peak-Hour Expressway Congestion', which does not convey the essence of his argument about the equilibrium between private and public transport. I have already noted that the 'Down's Law' for urban motorways has become well-established and I will give further examples later.

Indeed, Downs himself seems to have recanted somewhat. In Downs (1979, p.360) he argues as follows:

> 'Some city planners have mistakenly concluded they could reduce surface congestion by building subways or other costly

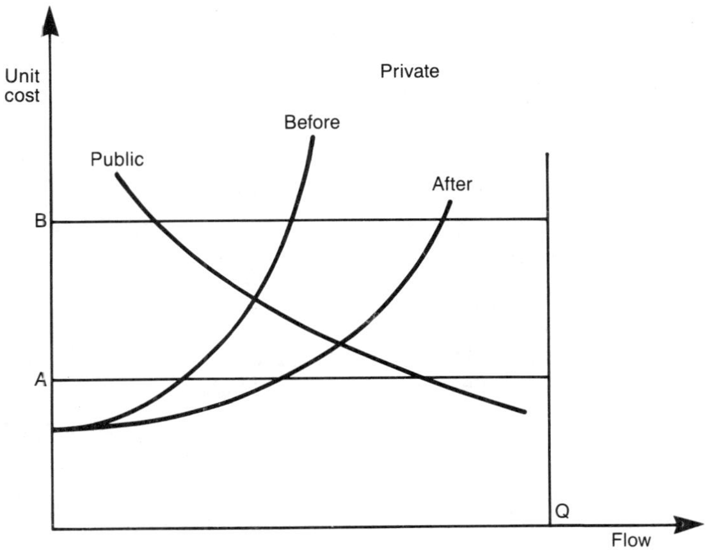

Figure 7.4 *The cost/flow equilibrium between private and public transport*

> off-street transit systems. Any space these systems initially free up on surface streets, however, soon fills with more cars and trucks as ownership and use expand.'

He quotes Owen (1972 pp.15–21) in support of this comment. This point is equivalent, however, to saying that the cost curve of segregated-track public transport systems is flat, as in Knight's original example of a broad road (Figure 6.2), rather than downward-sloping as in Figure 7.1.

The operation of the Downs-Thomson paradox, or the perverse reaction of the transport system operating in congested conditions, can be seen most simply in the above diagram, Figure 7.4.

The initial point of equilibrium is where the two costs are the same at A for the given total Q flow split between the private and public routes. If the road capacity is increased, then the cost curve for private transport is lowered. There is a modal shift from public to private transport so that the flow on the private transport increases whereas the flow on the public transport decreases. The new cost equilibrium point after the road capacity increase is therefore at the higher cost B, with the sum of the flows Q on public and private transport equal in the two before and after cases.

It will be seen that the reason the equilibrium costs increase after the road capacity is increased is that the decreasing public transport flows cause an increase in public transport unit costs. Thus, unlike the Knight example of two competing roads where the equilibrium costs remain the same until the narrow road can take all the traffic, in this case the equilibrium costs worsen until the car mode can take all the traffic.

Thus if road capacity is increased when it is impossible for everyone who wished to do so to travel by car, road costs at equilibrium increase. This is the Downs-Thomson paradox, or the perverse reaction of the transport system operating in congested conditions.

Thomson is conjoined to the paradox because he also recognised the phenomenon. In his book *Great Cities and their Traffic* (Thomson, 1977, p.165), in discussing his archetypal strong centre strategy, he put it this way:

> 'If the decision to use public or private transport is left to the free choice of the individual commuter, an equilibrium will be reached in which the overall attractiveness of the two systems is about equal, because if one is faster, cheaper and more agreeable than the other there will be a shift of passengers to it, rendering it more crowded while the other becomes less so, until a position is reached where no-one on either system thinks there is any advantage in changing to the other.
>
> 'Of course one must assume that those who still continue to choose private transport still find it preferable, despite the congestion and parking problems, and similarly those who choose public transport (but who could choose private transport if they wished) find it preferable, despite overcrowding, waiting and so on; but if a large number of travellers do have the choice and divide themselves between the two systems, one may infer that, in a rough and ready sense, the overall cost and quality of service is much the same on each system.
>
> 'Hence we derive one of the golden rules of urban transport: the quality of peak-hour travel by car tends to equal that of

> public transport . . . The important conclusion to be drawn is that all efforts to improve peak-hour travel by car will fail unless public transport is also improved.'

Thomson then went on to define the paradox in the following manner:

> 'Unhappily, attempts to improve traffic conditions by providing more capacity (on the roads) can lead to a deterioration in public transport, if, by drawing paying customers away from the latter, they force an increase in fares and a reduction in service. In this case, the balance of traffic may shift from public to private transport until a new equilibrium is reached where each system is of lower quality than before.'

Interestingly, Thomson did not define London as one of the strong centres in this classification; this grouping was essentially of those who were constructing major new facilities to enhance the dominance of the centre and to cope with projected traffic demands.

London was classified, contrariwise, as a strong centre having a traffic restraint strategy in which the potential demand for travel was not necessarily to be catered for, as we have already discussed. The paradox is, however, just as relevant to London as to any other city.

This equilibrium has also been recognised by Suchorzewski (1973) who stated it as follows.

> 'The volume of vehicular traffic will reach such a level that the speed of a motor vehicle in over-congested areas drops to the lowest acceptable level . . . The value of the lowest accepted speed—which can be called the critical speed—depends upon several factors; first upon the efficiency of public transportation. The more efficient and, in particular, the more rapid the public transportation, the higher the critical speed. This observation is essential for the further consideration of the search for an optimal system of urban transportation in the form of a so-called "balanced integrated urban transport system".'

Suchorzewski (1976) amplified this argument further, in particular distinguishing between road public transport and segregated track public transport, as Downs did, and again in Olszewski and Suchorzewski (1987) in which the access problems of parking are stressed.

Thus Suchorzewski did not recognise the perverse reaction of traffic speeds to increases in road capacity, but did recognise that it is public transport that has to be improved if traffic speeds are to be increased.

The first piece of work in London to use the fact that the public transport cost curves are downward-sloping was a pioneering study by the Labour administration of 1973–77. In the GLC Minutes of the meeting of 29 March 1977, we find (GLC, 1977, pp.41–55) a joint report from the Transport Committee and the Policy Resources Committee entitled *Transport in London—a Longer-Range View*. This was pioneering in the sense that it set out very clearly a set of policy options based on two extreme positions—called a public transport emphasis and a roads emphasis—and based on four different levels of spending over a 15-year period. Its aim was to set out the two kinds of London that the two different extremes of policy would be likely to achieve if given free reign, and thus to enable Londoners to express a choice between them.

Whilst it can be argued that this was also the aim of the LTS in 1968, the techniques used then, as discussed in Chapter Five, were in fact unable to do this adequately, particularly for public transport.

This was the first time in London then that a council had (a) considered such a wide range of possible capital and revenue spending options and (b) considered such policies from such extreme positions. It had, indeed, taken a considerable amount of argument within the administration, both professionally and politically, before the final report was presented to the council, in its dying days in 1977.

The report does not say what technique was used in the assessment of its policy options, but it was the SCENARIO model. Since this is essentially a short-term model, it is somewhat inappropriate for assessing the long-term implications of policy options.

The most important criterion in selecting public transport policies was that, for a given level of financial support for the public transport operator, LT, the number of passenger-miles it carried had to be a maximum. Strictly speaking, this was a marginal criterion to enable LT to judge the value of changes in fares and services policies (LTE, 1977), but here it was being used as an average criterion.

Passenger-miles maximization within a cost constraint is a marginal output measure to try to ensure that money resources are used most efficiently by the operator. It does not include the time resources of the passenger; that is, his or her walking and waiting time.

The key advance was that the different levels of expenditure clearly revealed the effect of the economies of scale, even if the result was partially obscured by the fact that the operations as a whole were reported rather than that of services at different intensities of operation.

There was no such optimum criterion for the roads since there was no measure available of the amount of traffic which would use a given capacity of the road system. It was impossible, therefore, to estimate the value of expenditure on roads except by discussion of the comparison with existing programmes of maintenance and capital expenditure.

After a brief discussion of the future trends expected in population, employment, incomes and car ownership, the report says:

> 'If no changes were made in the road network, the level of public transport services, the degree of traffic restraint, or the relative costs of travel as between the public and private modes then, based on the above projections of population, employment and car ownership, the following trends would apply:
>
> (1) a fairly constant total travel demand, with the fall due to declining population and employment being offset by increasing car ownership and use;
>
> (2) a substantial fall in work journeys to central and inner London;
>
> (3) some increase in journeys for other purposes, particularly in outer London and in off-peak hours;
>
> (4) an overall decline in public transport use, mainly on the buses; and
>
> (5) slow general growth in traffic on main roads, particularly in outer London.'
>
> GLC (1977, p.42)

We may also note a paragraph on efficiency of transport policy. The GLC was charged in the 1969 Act giving it control over LT to ensure that the public transport system was operated efficiently.

> 'It is also of importance to see whether the public transport system is run at a level where the benefits of scale become possible, both in terms of efficient operational units and in terms of the reduced waiting times associated with high service levels.'
>
> GLC (1977, p.43)

After a discussion of the financial responsibilities of the GLC, which did not include the trunk roads, mostly in outer London, or the BR rail system, and the cuts in its programme from £247 million in 1976–77 (roads £93 million, public transport (p.t.) £154 million) to £204M in 1977–78 (roads £82 million, p.t. £122 million) (1975 prices), the four options of expenditure were laid out:

> '(1) present government intentions to limit expenditure continuing throughout the 15-year period—the crisis level.
>
> (2) present reductions applying up to 1981, thereafter growth in line with the national economy (GDP), but London's share declining at the same rate as its population—the pessimistic level.
>
> (3) some easing of government policies before 1981, expenditure on transport restored to its pre-cut levels in the mid-1980s and growth in line with GDP; the older cities' share of expenditure declining less rapidly than population—the moderate recovery level.
>
> (4) considerable easing of government policies by 1981, expenditure increasing slightly above pre-cut levels in the early 1980s and growth in line with GDP; London's share falling less rapidly than its population—the full recovery level.'

This gave the estimates shown in Table 7.1 of levels of expenditure over the 15 years.

At the crisis level, roads expenditure was £1 billion, whereas p.t. expenditure was £1.5 billion. Two options for p.t. were calculated, a passenger–miles maximizing (PMM), high fares option and a

Table 7.1 GLC transport expenditure 1977–78 to 1991–92 (1975 prices)

Crisis	£2.5 billion
Pessimistic	£3 billion
Moderate recovery	£3.5 billion
Full recovery	£4 billion

Table 7.2 LT Service levels, fares and passenger miles—crisis level

		PMM	*Poor services*
Service level	– bus	80	65
	– underground	110	100
Fares		130	110
Passenger miles	– bus	72	71
	– underground	115	111
	– total	93	90

moderate real fares increase, poor services option. According to the model, these gave the service levels and passenger miles in 1991–92 as compared to an index of 100 in 1976–77 as shown in Table 7.2.

(The underground increase in passenger miles stems from an expected increase in tourism, the opening of the Piccadilly Line extension to Heathrow airport, and the first stage of the Fleet Line (since renamed the Jubilee Line)).

It is noted in discussing these results that a successful implementation of the Council's restraint target of a one-third reduction in car traffic in central London, together with additional restraint in inner London and in the 29 strategic centres of London, could have substantial benefits, especially if the revenues were to be used to support the Council's transport programme.

At the pessimistic level, roads expenditure was £1.2 billion, whereas p.t. expenditure was £1.8 billion. Two options for p.t. were again calculated, a passenger-miles maximizing, high fares option and a constant real fares, poor services option. According to the

Table 7.3 LT Service levels, fares and passenger miles—pessimistic level

		PMM	*Constant fares*
Service level	– bus	85	70
	– underground	110	100
Fares		117	100
Passenger miles	– bus	76	75
	– underground	116	114
	– total	96	94

model, these gave the service levels and passenger miles in 1991–92 as compared to an index of 100 in 1976–77 as shown in Table 7.3.

At the moderate recovery level, two options were explored, a roads emphasis and a p.t emphasis. In the former, roads expenditure was £1.7 billion, whereas p.t. expenditure was £1.8 billion, the same as in the pessimistic level. To make comparison easier, the same p.t. programme for passenger-miles maximizing was used as in that level, but no account was taken of the slight increase in fares that might be necessary to take account of the decline in patronage because of the additional car use that might be generated.

In the p.t emphasis, the roads programme was kept to that of the pessimistic level at £1.2 billion, although there is a slight shift from capital to traffic management to enable bus priority measures to be implemented, as part of a passenger-miles maximizing option developed for p.t. with a total expenditure of £2.3 billion.

According to the model, these gave the service levels and passenger miles in 1991–92 as compared to an index of 100 in 1976–77 as shown in Table 7.4.

At the full recovery level, the options were not explored in detail, but it was suggested that a middle course of taking the roads element of the roads programme at the moderate recovery level at £1.7 billion, together with the p.t. element of the p.t programme at that level at £2.3 billion could be taken; that is, amounting to the full £4 billion programme.

It was finally suggested that a programme somewhere between the two extremes of the moderate recovery level should be planned for.

Table 7.4 LT Service levels, fares and passenger miles—moderate recovery level

		Roads emphasis	*P.T. emphasis*
Service level	– bus	85	90
	– underground	110	115
Fares		117	96
Passenger miles	– bus	76	82
	– underground	116	120
	– total	96	101

It is interesting to note that the policy options generally take into account the passenger-miles maximizing criterion developed by LT as a guide to the allocation of resources. This tends to give more emphasis to service levels than to fares, since measurements of passenger behaviour showed they were about twice as sensitive to the former as to the latter. There were, however, various policy options which attempted to minimize fares, as this was thought by the Labour administration to be important for poorer passengers, even though less efficient in its use of resources.

The effect of the economies of scale can be seen by examining the total costs of the operation envisaged in 1991–92 as compared to the output passenger-miles carried in Table 7.5.

The use of 1976 out-turn prices as a base whereas the forecasts are given in November 1975 prices will increase row (a) and thus cause an overestimate of the marginal costs on the buses and an underestimate of the marginal cost on the railway.

It can be seen that the higher the level of passenger-miles carried, the higher the passenger-miles per pound, or the lower the cost per passenger-mile, though the amount of variation was not substantial in the options studied. It is also clear that the subsidy to operating costs had to increase substantially if the gain in passenger-miles per pound was to be achieved on the buses, whereas on the rail system the subsidy was actually reduced—the gain in passenger-miles comes from the opening of new lines and the increase in tourism, as already noted.

Table 7.5 LT operating costs and passenger-miles (PM), 1991–92 forecast (Passenger-miles maximization) (November 1975 prices)

	1976 (out-turn prices)	*crisis*	*pessimistic*	*moderate recovery*
Bus				
Operating costs £M	209	172	179	186
Receipts £M	128	102	94	79
Subsidy £M	81	70	85	107
% subsidy	39	41	47	58
PM (B)	2.98	2.15	2.26	2.44
PM/£	14.3	12.6	12.7	13.1
p/PM	7.0	8.0	7.9	7.6
Marginal cost £M		–37	–30	–23
Marginal PM (B)		–0.83	–0.72	–0.54
Marginal p/PM		4.5	4.2	4.3
Marginal/average p/PM		0.57	0.53	0.57
Rail				
Operating costs £M	162	162	162	164
Receipts £M	122	165	153	132
Subsidy £M	40	–3	9	32
% subsidy	25	–2	6	20
PM (B)	2.71	3.12	3.14	3.25
PM/£	16.7	19.3	19.4	19.8
p/PM	6.0	5.2	5.2	5.1
Marginal cost £M		0	0	2
Marginal PM (B)		0.41	0.43	0.54
Marginal p/PM		0	0	0.37
Marginal/average p/PM		0	0	0.07

Adapted from GLC (1977, Appendix B)

The increasing subsidy was, therefore, given mainly to the bus system rather than to the rail system, despite the fact that the bus system was more expensive per passenger-mile, both in average and in marginal costs.

The marginal cost per passenger-mile for bus is thus probably about half of the average cost, whilst for rail the marginal cost is a much smaller proportion of the average cost. This is not surprising. Once a rail system is established, it can relatively easily carry more passengers for relatively little extra cost if operating well below capacity, thus reducing unit costs per passenger-mile carried. The costs of a bus system per passenger-mile carried, on the other hand, are relatively insensitive to the number of passengers, since extra capacity has to be provided. In other words, the economies of scale are much greater on a rail system than on a bus system.

Similar results were found by Train (1977) in an analysis of the San Francisco Bay Area Rapid Transit (BART) rail system as compared to the Alameda-Contra Costa (A.C.) Transit bus system. He found that the rail system average costs were below those of the bus system, and that the rail system marginal costs were probably substantially below those of the bus system.

The passenger-miles per pound of total operating costs give a measure of the average efficiency of the use of resources. What is required to establish the level of subsidy is the passenger-miles gained or lost per pound of marginal cost, either passenger-miles gained per pound extra spent or passenger-miles lost per pound saved.

When the GLC was presented in this period (1973–77) with the analysis of LT bus projects to demonstrate which ones gave the maximum passenger-miles per pound, it was found that the pass-mark, for projects worth implementing at the predetermined level of subsidy, was about twice the level of the average passenger-miles achieved per pound of operating costs. Rail capital projects were found to give a much higher passenger-miles per pound value, but owing to the government restriction on capital spending and to the predilection of the Labour administration for revenue support rather than capital investment very few capital projects were approved.

This work demonstrated that, if LT were to operate to maximize its efficiency in the use of resources, then a subsidy to bus operating costs of about 50 percent was needed at the current level of demand. This was thus consistent with the long-term strategy's marginal costs as a proportion of average costs. Where the long-term strategy went wrong was to ignore the large gains to be made in passenger-miles by concentrating the subsidy more on the rail than the bus system. It will also be remembered that the GLC had no power over the BR rail system, so did not incorporate any suggestions about the development of the BR system in its strategy.

The long-term strategy was dependent on the level of subsidy to achieve the gains in passenger-miles. Such a high level was never likely to be approved at that time, as the main argument was in fact about the level of spending by government, not about its efficiency.

The long-term strategy was, therefore, short-lived, for the incoming Conservative administration in 1977 promptly shelved the proposals and the techniques that underlay them. It began to concentrate, instead, on attempting to improve the efficiency of operations of LT and to reduce its costs.

I left the GLC in 1978, to go back to my old college and write up my work on car ownership with a grant from the Nuffield Foundation (Mogridge, 1983). Within a few months of my departure, all my colleagues in the Economics Section had also left; Lewis to join the USA government in Washington, Cheung and Garden to join the Dutch government in the Hague, Goodwin to join and ultimately to become the Director of the Transport Studies Unit at Oxford, and Fisher and Godier to join public transport executives.

Goodwin and myself had been asked, before we left the GLC, together with Collins of LT, to join an international working party under Webster of the TRRL to investigate the factors affecting public transport demand (TRRL, 1980). After a first meeting at the GLC, subsequent meetings were held in Munich, Stockholm and Paris, which enabled the members of the group to see the developments in these cities at first hand.

We were each given a responsibility for summarizing the evidence presented to us in a particular area. Mine was the interaction of transport and land use, which was particularly important to London. The forecasts of future travel made in 1966 by the LTS and 1977 by the GLC had been so different partly because of the difference in the forecast estimates of the spatial location of the population, employment and income. My own interest in this subject had begun with Mogridge (1968) and (1973), but the exposure to the evidence presented to the members of the working party, and the subsequent invitation to me to work at the Institute de Recherche des Transports in Paris for a period in 1979–80 (Mogridge, 1980b), enabled me to gain a new perspective on the subject.

It will be recalled from the beginning of the previous chapter that Smeed (1961) started his paper on *The Traffic Problem in Towns* with a discussion of the use of space in the city. In fact, before that, he had discussed the way in which cities are spreading out with time, and I will return to that problem in Chapter Nine.

Firstly, however, I want to note that, at the conclusion of the working party on public transport demand, a further working party was set up, again under Webster at the TRRL, to investigate how the then current land use/transport interaction models compared with each other.

Whilst I was not a member of this new working party, I was asked to undertake a comparison of the data available in London and Paris on transport facilities and travel patterns (Mogridge, 1983b, 1984). It was during this work that, in using Thomson's (1977) book describing the two cities, I realised that his ideas described above could, perhaps, be substantiated, on the one hand with the data from the survey of random journeys in central London by the Transport Studies Group coupled with additional journeys on the rail system (Ferguson, 1985), and on the other with the data from the GLTS on actual journeys. These analyses have already been given in Chapters Two-Four.

I first described our attempts at substantiating Thomson's ideas in a lecture at University College London on 17 October, 1985 (Mogridge, 1986a); the title of that lecture forms the sub-title to this book. Whereas such lectures are usually taken from—or used as a basis for—a more rigorous presentation in an academic paper (Mogridge *et al*,1987), (by which time I had become aware of Downs' contribution), my lecture in its written form received the distinction of a critical analysis by Bly, Johnston and Webster (1987), and subsequent to our reply (Mogridge and Holden (1987)) to a further analysis (Johnston (1988)).

It is on Johnston's (1988, p.7) use of the multinomial logit modal split model that I wish to focus at this point of the argument.

Johnston points out that the multinomial logit model is a special case of the general random utility choice model, and its derivation by utility maximization methods is given by Domencich and MacFadden (1975) in the context of a general theory of transport modelling.

The key concept is that, for a particular traveller wishing to make a particular journey, there will be a random utility offered to him by a particular mode, drawn from a distribution of utilities according to his or her perception of the advantages and disadvantages of that mode of travel for that journey. Other modes will offer different average utilities for the same journey. The traveller chooses that mode which offers him or her the highest utility out of those available. This is tantamount to saying the traveller chooses the best mode as he perceives it.

The important result of these choices is that the achieved utility on a given mode is not the same as the offered utility, since travellers are more likely to reject low utilities because other modes are more likely to offer a higher. Moreover, the average achieved utility on all modes actually used for a given journey will be the same, together with the distributions of those utilities on each mode, provided all travellers have had a free choice between the same modes.

This is, in fact, no more than a rigorous mathematical statement of the concepts put forward by Downs and Thomson, and generalizing Wardrop's first criterion, but it allows one to go on to explore some consequences. Chief among these is the fact that improvements to the dominant mode must be made if the achieved utility is to be maximized. This can be seen very simply in the presentation of Figure 7.4.

Johnston goes on to point out that it is the cost/benefit ratio that determines whether a mode should be improved or not. Improvements to the dominant mode, he argues, may not give the greatest improvement in terms of the increased mean achieved utility for a given expenditure.

Although he accepts that the offered utility of the public transport modes would fall if the loss of revenue caused services to deteriorate or become more expensive, as both Downs and Thomson argue, Johnston is unwilling to accept that where public transport is the dominant mode it should be the mode to be improved. His argument is expressly about the dominance of public transport, and especially rail commuting to central London in the peak-hour. He argues instead that:

> 'Except to the extent that they attract existing car travellers and so reduce traffic congestion, rail improvements provide benefits confined to those who can travel by rail on that route at the time that the improvement applies. Benefits from a road improvement are likely to be more widespread, applying not only, or even primarily, to peak-hour commuters to the central area making journeys along the obvious line-haul route, but also to a potentially large number of non-central car trips which might use the improved road as part of the route, to off-peak journeys, and to freight hauliers' journeys for the carriage of goods.'

He recognizes that:

> 'New urban road space is expensive (though traffic management measures can be relatively cheap), and it is not possible to generalize about which modal improvements are likely to be most cost-effective.'

He admits:

> 'This needs a proper study covering all the adaptive responses for each particular case.'
>
> Johnston (1988, p.9)

This, of course, is precisely what the standard transportation planning model is set up to do, provided it has a multinomial logit model at its heart to model modal split (or other similar modal split models), and provided it has the correct response to increased flows, not only for roads but also for public transport modes. I will return to discuss the advances made in the London strategic model in the 1980s in the next chapter.

The dominant mode issue can be clarified by examining the modal split of journeys by the distance of the origin or destination from the town centre. The first project of this type in the UK was that by Outram (1972), who analysed the 1966 journey-to-work data for a large number of British cities. These data, as already noted, were the first to be geo-coded by the census, and are a 10 percent sample of work journeys.

The analyses I did of the modal splits by distance from the centre for the journey-to-work on 1971 London data (Mogridge, 1978, p.62; 1979, Figure 3) and 1976 Paris data (Mogridge, 1980c, Figures 13b and 14b) were presented together in the working party report (TRRL, 1980, p.163, Figure 9.2). The London data were from the 1971 census normal journey-to-work; the Paris data were from the 1976 transportation survey first actual journey-to-work. My final report to the Institute de Recherche des Transports (Mogridge, 1980b, Figure 10.2) also gave the results for London and Paris together, and it has subsequently been used in Mogridge (1983b, Figure 18; 1986b, Figure 8); the Paris curves were used in Fradat *et al* (1984, p.23). They are shown in Figure 7.5.

The description of these curves in (TRRL, 1980) is as follows:

> 'Trips by car become predominant for journeys from residences in the outer suburbs, while trips on foot are predominant for those living in the centre. In Paris, rail is also heavily used for trips from homes close to the centre, possibly because stations on the Metro system are relatively closely spaced, so encouraging quite short journeys by rail, while in London bus takes a larger share of such journeys.
>
> 'In both cities, rail is predominant for trips to workplaces in the centre: both cities have very extensive rail systems, of course, with underground rail being used for much of the commuter travel in from the suburbs, while the surface network caters for longer-distance commuter travel.
>
> 'Again, car is predominant for travel to work places in the outer suburbs with car use being higher, and bus lower, in Paris than in London. This is largely a reflection of the higher level of car ownership in Paris than in London (the Paris data are 5 years later than the London data), but London's bus network is also more extensive than that in Paris.'
>
> (TRRL, 1980, p.162)

It can be clearly seen that for London and Paris, public transport modes are dominant for work journeys near the city centre, whereas as the distance of the residence and workplace from the centre increases then so the dominance of the car replaces that of the public transport modes. The decision on the appropriate transport investments must therefore take this into account.

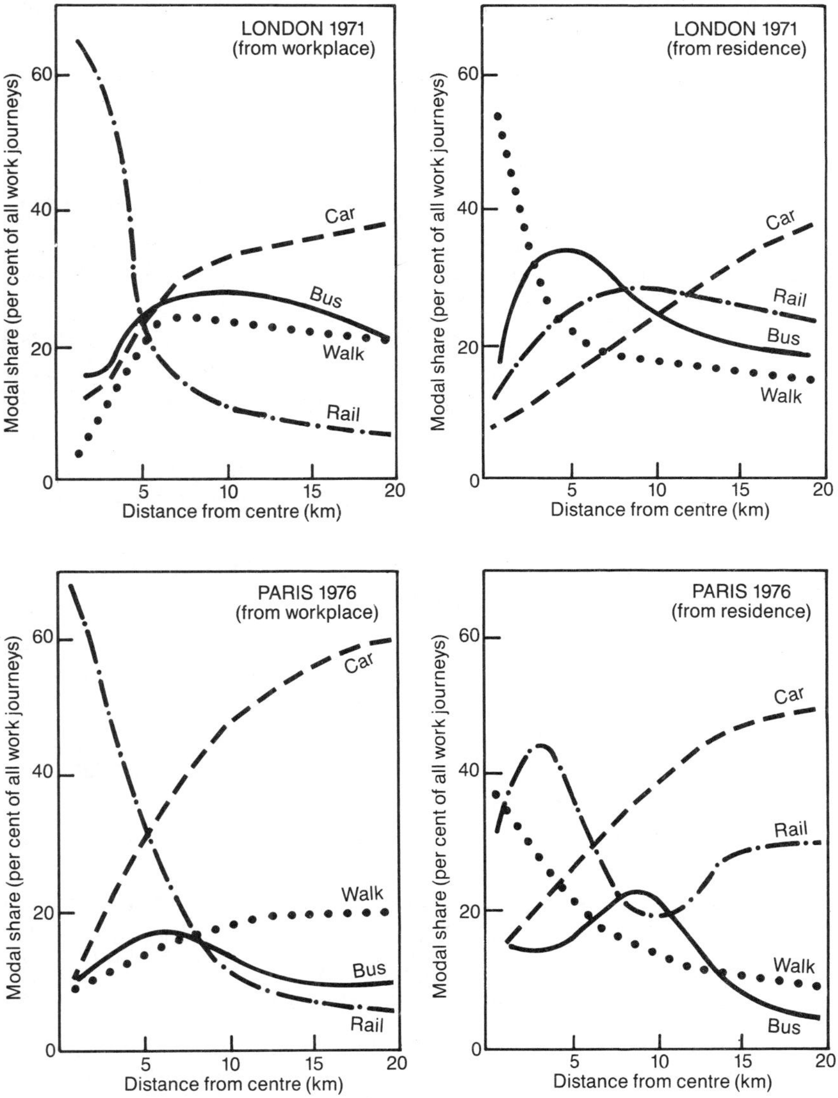

Figure 7.5 Modal split by distance from the centre for the journey-to-work, London 1971 and Paris 1976

The 1971 and 1981 data for modal split by distance of the residence from the centre of the city for the census journey-to-work has been extensively studied for British cities by Dasgupta *et al*(1989), relating the mode choice to sex differences, socio-economic group and household car ownership.

The validity of the standard transportation model relies on an assumption that travellers actually do attempt to choose the mode that is best for them and that they will switch between modes during the course of time in order to produce a dynamic equilibrium, as Downs (1962) suggests in his terms 'explorers' and 'sheep'.

Holden (1989) points out that Wardrop's (1952) two principles are static criteria; a dynamic criterion is implied where drivers search for the minimum route. Given a typical coefficient of variation of 0.20 on an individual route in urban areas (Smeed and Jeffcoate, 1971), it is clear that drivers are unlikely to be able to discriminate between two routes of similar trip times, especially given the possibility of unpredictable traffic jams.

Moreover, some drivers must always be on routes which are not a minimum if they are searching for routes which are. If such searching takes place, then the average trip time will tend to a minimum. It is obviously essential for the stability of any network parameters, flow, speed and concentration, for such an equilibrium to exist—they could not otherwise be defined.

Beardwood (1989) has described a neat demonstration of the process of equilibriation in a two-route situation between a zone-pair, one route being high fixed cost and zero variable cost and the other low fixed cost and high variable cost, representing public transport and car respectively. With 100 travellers choosing from Weibull distributions of offered costs at random, and a multinomial logit modal split model, as suggested by Johnston (1988), she could show the development through time of the costs and flows, and thus how equilibrium is approached. This reveals quite clearly the difficulty of measuring the situation to verify that equilibrium occurs.

The crucial test of this equilibrium theory is, of course, to see if travellers actually do switch between modes during the course of time. One early indication that they might do so was given in the surveys of 'Private Motoring in England and Wales' reported in Gray (1969). In these surveys approximately 3,000 households were

Table 7.6 Percentage of household cars used for work journeys on a given number of days in a week, Greater London 1961–64

	Oct 1961	*Apr 1962*	*Jul 1962*	*Oct 1962*	*Jan 1963*	*Apr 1963*	*Jul 1963*	*Oct 1963*	*Jan 1964*
5 or more days	34	38	35	41	29	48	42	45	38
3 or 4 days	12	13	12	13	10	14	6	10	12
1 or 2 days	12	10	14	7	13	11	14	7	10
Not at all	42	39	39	39	48	27	38	38	40
Total	100	100	100	100	100	100	100	100	100

interviewed in each of nine periods between October 1961 and January 1964, with about 1,000 household cars in each survey. They were the precursors to the National Travel Surveys.

Table 7.6 repeats Gray's Table 3.11 for Greater London to show the extent to which cars were used for the journey to and from work over the week of the survey.

The data for January 1963 is affected by the severe winter. It can be seen that not all cars were used for work journeys on five or more days a week, implying some mode switching (as well as days off work through illness and so on).

An analysis of mode switching can be carried out using the London Panel Surveys of 1982–84, which examined the travel behaviour of approximately 2,000 individuals over the course of a seven-day period. These surveys were repeated with essentially the same individuals every six months. The results of Terzis and Mogridge (1988) are summarized in the following paragraphs.

We take the work journey, as this is a repeated journey between the same origin and destination each day. There is a small definitional problem here, as the Panel Surveys include both education trips, and trips in the course of work (on employer's business), in the definition of work trips.

In Table 7.7 we give the numbers and percentages of workers who only travel on one mode for all the trips to work in the survey week (uni-modal) and those who travelled by a mixture of modes (modal-mix). It can be seen that in each survey, some 15 percent of

Table 7.7 Numbers and percentages of respondents by the different kinds of mode-choice behaviour over a week-long period for the four panel data sets

	Autumn 82	*Spring 83*	*Autumn 83*	*Autumn 84*
Uni-modal	906 (86.2%)	1177 (86.2%)	929 (86.2%)	1078 (84.9%)
Modal-mix	157 (14.8%)	204 (14.8%)	161 (14.8%)	192 (15.1%)
Total	1063	1381	1090	1270

[1] *The uni-modal category includes respondents who travelled by a single mode over seven days.*

Table 7.8 Modal-choice behaviour for respondents who work, full- and part-time[1]

	Full-time	*Part-time*
Uni-modal[2]	85.4%	84.5%
Modal-mix	14.6%	15.5%
Total	787 (100%)	122 (100%)

[1] *Total part- and full-time workers 909, total sample size 1090 respondents in autumn 1983.*

[2] *The uni-modal category includes respondents who travelled by a single mode over seven days.*

respondents travelled by a mixture of modes, and thus revealed that it is possible to travel by different methods on different days.

In Table 7.8 we show that this proportion of 15 percent is not subject to the difference between those who work full time and those who work part time.

In Table 7.9 we show that switching occurs between all possible mode combinations: bus-car, bus-rail, car-rail and others. It is worth noting the substantial increase in the proportion of switches between bus and rail in Autumn 1983 compared to a year earlier, after the introduction of the composite pass in May 1983. It did not, however, seem to affect the switching between car and bus or between car and rail.

Table 7.9 Percentage of respondents by the various kinds of mode-mix behaviour

	Autumn 1982	*Autumn 1983*
Bus–car	17.2%	17.4%
Bus–rail	7.0%	15.5%
Bus–other[1]	10.8%	12.6%
Car–other	19.8%	14.1%
Car–rail	22.9%	22.4%
Rail–other	10.8%	9.3%
Other[2]	11.5%	8.7%

[1] *Other mode, includes walk, cycling etc.*

[2] *It includes people who were observed to have a choice-set of more than two modes.*

Table 7.10 Modal-choice behaviour, comparisons over time

	Spring 1983			*Autumn 1983*		
	Modal-mix	*Uni-modal*	*Total A82*	*Modal-mix*	*Uni-modal*	*Total A82*
Autumn 1982						
Modal-mix	24	57	81	23	35	58
Uni-modal[1]	60	353	413	37	244	281
Total S83	84	410	494			
Total A83				60	279	339

[1] *The uni-modal category includes respondents who travelled by a single mode over seven days.*

In the final table from this work, we quote in Table 7.10 the numbers of respondents who were in two successive surveys, six months and a year apart, respectively, to show whether this switching is constant or not. Rather unexpectedly, we find that only a small proportion of the people who switched in the first of the two surveys were still switching in the second survey; but that a different set of respondents were now switching, so that the proportion who were switching in the second survey was very similar in total to that of the first. The numbers of respondents is much smaller in these tables than in the earlier, because of drop-out (or attrition).

The proportion switching is 81 or 84 in 494 (16.4 or 17.0 percent) in the first pair of surveys and 58 or 60 in 339 (17.1 or 17.7 percent) in the second pair, but the proportion switching in both is only 24 in 494 (4.9 percent) in the first pair and 23 in 339 (6.8 percent) in the second pair. One can only but presume that a much larger proportion than 15 percent of journeys are potential mode-switching journeys.

It can thus be seen that it is obviously possible for an equilibrium between modes to be established, because a large proportion of travellers switch between modes in the course of time, at least in London.

8 *Public Transport*

Meanwhile at the GLC from 1978 onwards, further work had proceeded to resolve some of the remaining problem areas in the STEM network model. In Blase (1980) some of these improvements were listed:

> it was realized during the second GLTS calibration of the 1971 data that junctions were important, especially in the central area, and that speed/flow curves on links were inadequate to deal with crossing traffic at junctions. Initially, from 1975, a rather simple formula was used to incorporate delays at junctions, but from 1978 the standard formula devised by Webster (1958) was used. Whilst this considerably increased computing time, it was realized that this was worthwhile.
>
> the value of time for car and public transport users was changed to a common value when evaluating benefits, rather than using separate values to represent the average user of each existing mode as in the calibration and forecasting stages.
>
> the car operating cost was changed to incorporate link time as well as link distance.
>
> the deterrence functions were estimated from a computer-based method, rather than the relevant parameters being estimated manually from the output and then being fed back into the computer.

The drawbacks to the STEM still remaining in 1980 were:

> the road model still operated at a district level (250 districts) whereas the public transport model operated at a zonal level (1,000 zones), giving a discrepancy of scale and difficulties at the modal split stage. This differentiation was and still is necessary; the shorter bus trips require a zonal model whilst

the slightly longer car trips are adequately modelled at district level, thus minimizing computer costs.

the calibration areas for the deterrence functions were unchanged, and thus suffered from the drawback that from given origins, trip numbers could increase with trip costs, in contravention of Beckman *et al*'s second criterion, thus giving a possible spurious traffic equilibrium. The risk of this was, however, thought to be slight, and unavoidable.

public transport fares were still estimated from the route distance between zones rather than being incorporated as a zone-to-zone matrix. It was not until 1984 that this change was finally implemented, after the new GLC fare zone system had been established by the Labour administration.

access costs for cars, particularly to the central area, were still unresolved.

assignment was still incremental and not iterative, and used four iterations before stopping. This means that an inefficient 'capacity restraint' procedure was used rather than the correct, efficient procedure for finding a true equilibrium devised by Frank and Wolfe (1956). This latter procedure was described by Van Vliet and Dow (1979) and essentially ensures that the flow is efficiently partitioned between links on successive iterations to ensure that Wardrop's first criterion is met. It was finally implemented into the STEM in 1983. Eleven iterations were used, with about 17 percent of flow being assigned on the first iteration and approximately equal proportions of flow thereafter.

By the time of the abolition of the GLC, therefore, the STEM—now renamed the GLTS model for the 1981 calibration—was becoming a useful tool to evaluate transportation proposals, though still working on average costs or perceived costs rather than marginal costs.

The GLC Plan *Public transport in London: the next 10 years* (GLC, 1985) was produced in the Council's last few months before abolition, and after the national government had taken back control over LT in 1984 and renamed it London Regional Transport (LRT). The director of the GLC study team, Niblett, has subsequently become BR Network SouthEast Strategic Planning Manager.

The GLTS model was extensively used in evaluating the proposals in this plan.

By this time, the GLC had implemented its zonal fares system in May 1983, and reduced fares, but the general planning outlook remained somewhat depressing. Population was expected to remain constant or perhaps decline slowly by about 1 percent per year, much slower than in the 1970s; whereas employment was expected to remain relatively stable and unemployment—which had reached a peak in the early 1980s—to decline. Car ownership was expected to continue to increase, reflecting increasing income. The Plan concludes this section on planning parameters by asserting:

> 'In total, the planning parameters could by themselves lead to a decline in public transport use of up to 1 percent to 2 percent a year but this would be much more than offset by the growth trends flowing from the public transport initiatives proposed in the Plan.'
>
> GLC (1985, p.17)

This Plan is remarkable for setting out clearly a number of guidelines and standards which the then GLC felt should be applied to the public transport system if it were to become a fully integrated system.

It is also notable for commenting that:

> 'On a few major corridors with very heavy flows of buses, it is possible that investment in a more capital-intensive system such as light rail could be justified by the improvements in speed, reliability and efficiency which could be achieved.'

It continues:

> 'To provide an attractive service the light rail would need to be given high priority over other traffic with exclusive running lanes wherever possible. Extensive traffic management would be required to achieve this and provide safe and convenient stops for passengers.'

It therefore concluded that:

> 'a study should be carried out by the GLC and LRT into the possibilities for a limited number of light rail services operating on-street in inner and central London.'
>
> GLC (1985, p.53)

This has subsequently begun, although the focus of attention has been on outer London in Croydon, using in part old rail lines.

The Plan gives detailed costs and benefits of a number of proposals for integration of LRT and BR rail lines. It points out that several current north London LT lines were taken over from BR and suggests that this process of integration should now be applied to other London lines. The technical difficulty that tube lines are too small a gauge to allow through running by BR stock can, of course, be overcome by running tube stock on the BR track, just as on the north London lines. Similarly, the difference in electrical systems can be overcome by dual-voltage engines.

A further study of one of the proposals in the London Rail Study (Barran, 1974), the re-opening of the Snow Hill connection between north and south London BR lines, had already been funded by the GLC Labour administration. This had shown:

> 'the major advantages of both improved access across the Central Area and reduced operating costs resulting from trains not having to turn round at central termini.'
>
> GLC (1985, p.37)

The Plan reported government approval of the scheme. The link (the Thameslink) was opened in May 1988 and has already proved so successful that the link is to be further developed to provide a higher level of service. I discuss this later.

On the same theme of cross-central London rail connections, the Plan continued by noting that the 1974 London Rail Study had examined the Cross-Rail scheme. It commented that:

> 'Analysis showed heavy use and large passenger benefits for both lines, and the Study recommended further examination. This has not been carried out, mainly because BR's finances have been restrained by the government.'

After abolition, the responsibility for running the GLTS model rested with the successor body, the London Residuary Body, for a while, before the Department of Transport assumed the responsibility and promptly placed it with private consultants in 1987.

A technical audit of the model was commissioned from the consultants (MVA, 1988) and a criticism of its accuracy in determining traffic flows was also commissioned from other consultants (Wootton Jeffreys, 1989), for whom I acted as an adviser.

We noted that, since the Frank-Wolfe procedure (to find the best partitioning fraction) is calculated over all links in the network, this may give rise to problems in the central area—where most of the congestion occurs—because it is a small part of the total area.

> 'The first assignment, for routes going through the central area, may not be on a suitable path, and the iterative process may not switch a high enough proportion of the flow to viable alternatives . . . The problem is ameliorated by starting from known speeds and flows (for junction delay calculations), as in the GLTS procedure.'
>
> Wootton Jeffreys (1989, paragraph 10.18)

We went on to note that:

> 'Research has shown that (on the 1971 model) a zonal-level assignment (of traffic flow) gave 18 percent more vehicle mileage than a district assignment (Anderton, 1978, Table 3.2 peak-hour). This is mainly due to centroid connector problems, and to intra-district traffic . . .
>
> [If] a single zero-length centroid connector were defined to the node closest to the centre of the district (then) unpublished results from the LTS Unit show that this method gives little difference between the total vehicle kilometres for zonal and district level assignments.'
>
> Wootton Jeffreys (1989, paragraphs 10.24–25)

This is because most car trips are very short and many are comparable in size to districts and thus not assigned to an inter-district network. This is not so much a problem in the central area, where both zones and districts are much smaller in area.

The Department of Transport and MVA decided in 1988 to

incorporate the multinomial logit model in the heart of the distribution and modal split process, which will enable the findings of Johnston (1988) to be applied and ensure that achieved costs are equal on competing modes where there is a capacity limitation. This has, however, not been done in time to affect any of the results reported later in this chapter.

But the arguments discussed in the last chapter about marginal costs have yet to be incorporated in the modelling. A further discussion about the history of these ideas may clarify the problem.

In 1952, Wardrop stated two criteria for route selection through a road network as already noted. The first is the basis for all transportation network route-finding procedures; the second is an optimization criterion for the network and has been almost totally neglected.

To recap, the second criterion says that:

> 'the travel time in a network will be a minimum'.

If we generalize this, we obtain the criterion that travel cost in a network will be a minimum, where cost implies money, time and effort, and we have a fixed pattern of trips to allocate to the network, where network implies all the possible methods of travel.

Van Vliet and Dow (1979), as well as describing the procedure for finding the equilibrium point for the first criterion, also investigated the procedure for finding the equilibrium point which ensured that the second criterion is met. They concluded, when testing their procedures on Leeds data, that a reduction of about 2 percent in total travel time could be achieved as compared with an equilibrium using the first criterion. It must be remembered, however, that this 2 percent refers to the metropolitan area; the proportion of time saved in the central area where congestion is greatest is much higher, though an estimate of this is not given by Van Vliet and Dow. Nor did Van Vliet and Dow find the optimum for travel as a whole, merely for travellers on the road network.

Essentially, what Wardrop's second criterion does is to focus on the marginal costs of car travel and not the average in order to obtain a system optimum, rather than a user optimum which is what the first criterion gives.

The procedure for optimizing modal split relies essentially on defining the marginal cost/flow relation. This has not yet been done in London. A system optimum cannot therefore be defined. There are, however, indications of the value of public transport subsidy in optimizing resources.

The work begun in TRRL (1980) eventually led to a major study analysing the effect of subsidies on bus services in the United Kingdom. Bly and Oldfield (1986a) produced a mathematical model which enables some of the questions about marginal cost to be tackled. This paper is a simplified version of a detailed mathematical model analysing the cost and benefits of subsidies in the Metropolitan Counties of the United Kingdom (TRRL, 1983). The aim of the study can be stated as follows:

> 'In order to judge whether or not the subsidy is wisely spent, it is necessary to decide to what extent (any) increase in passengers represents a benefit to the community which outweighs the cost of the subsidy.'
>
> Bly and Oldfield (1986a, p.200)

Bly and Oldfield continued:

> 'It is assumed that any increase in general subsidy can be used either to increase the amount of service offered, in terms of bus service frequency on each route or in terms of route coverage, or to reduce overall fare levels. Three components of benefit are considered:
>
> (i) The advantage in reduced fares or shorter travel times to passengers already using the service;
> (ii) The advantages to passengers newly attracted to bus travel because of the lower fares or better services;
> (iii) "External" benefits, including those to road traffic in general because modal transfer from car to bus reduces traffic congestion and therefore increases road speeds for all.
>
> In addition to the effect of subsidy on fare levels and passenger time spent waiting for the bus or walking to the route, the mathematical model also considers the impact of increased patronage on average bus occupancies, and the way these are likely to affect bus speed, bus cost and passenger waiting time.'
>
> Bly and Oldfield (1986a, p.201)

The difference from Mohring's (1972) study is that this study looks at the total network, rather than at a single route at a time.

The essential input is a constant elasticity of demand for travel with respect to generalized cost and the essential constraint is that the benefit must be maximized for a given subsidy so that both fares and service levels are adjusted as subsidy levels change.

Unlike Mohring, Bly and Oldfield claim that:

> 'In the long-term where all fixed costs can be assumed to respond to changes in service provision (there seem to be no genuine economies of scale in bus operations), the cost elasticity will be unity'
>
> Bly and Oldfield (1986a, p.204)

It should be remembered that this cost is the money cost of operating the service; it does not include the time cost of the passengers which, as Bly and Oldfield describe, will decrease as flow increases because of lower walking and waiting times.

They show that the average money cost per passenger journey should be equal to the average waiting cost and also to the average walking cost (since these two latter terms must be equal if the service is optimized) provided the second order effects are ignored (that is, the effects of bus occupancy and the external effects on road congestion etc.). The relevance of these calculations to the data on journey times presented in Chapter Three should be obvious.

They also show that the marginal rate of return on subsidy is greatest at the demand level where fares cover all operating costs, but is still positive, at a typical fares elasticity of 0.33, even when the total subsidy is at levels greater than the original total operating costs. Because the demand has increased, the level of subsidy per passenger is less than the previous fare at break-even, even at the point where subsidy covers all operating costs.

After the election of the Conservative government in 1983, with its manifesto promise to abolish the GLC, the Department of Transport began to prepare for its future responsibilities in strategic transport planning for London. The TRRL was asked to devise a new model which would incorporate the developing ideas about public transport, and in particular the relationship between fares, service levels and subsidy; but which would provide the following facilities, according to the brief:

> 'It is intended as a tool for strategic policy analysis, in that it provides a quick, easy and cheap facility for examining how travel patterns in the London area are likely to change in the future under a wide range of possible conditions. It is constructed in a way which makes it relatively easy to modify the inbuilt relationships if the user should wish to change particular assumptions, or to examine the sensitivity of the results to these assumptions, or to postulate additional mechanisms affecting various aspects of the model.
>
> 'The model is seen as a preliminary, "broad brush" method of assessing policies; it can be used as a rapid and easily adjustable filter for testing a broad range of possibilities and identifying those which seem worth subjecting to a more detailed, but more time-consuming, examination available from network-based land-use/transport models of the type provided by London Residuary Body's STEM, for example. It may also be useful in checking the output from these larger models, and in some applications it may be possible to use the model to formulate some of the input data for running the large model.'
>
> Bly and Oldfield (1986b)

The London Area Model (LAM) was the result. Like CRISTAL, it dropped the idea of a detailed network representation, but unlike the former, it was not a spatially symmetric model but rather a coarse spatial model, dividing London up into the central area, and then four quadrants for the Inner City, Inner and Outer London, and four quadrants for areas external to the GLTS area; making a total of 17 zones in all. These are shown in Figure 8.1.

I and my colleagues at University College were involved in helping to prepare the data for this particular zoning system by extracting the data for the TRRL from the 1981 and 1971 GLTS which we held.

The model can, therefore, only be used at the strategic level and not at a detailed local level. The key to the philosophy underlying the model is the response to change; it was not intended that the model should give a detailed reproduction of the cross-sectional data (i.e. at a moment in time) but rather that its response through time should agree with the various longitudinal studies which had been done, particularly those regarding elasticities with respect to fares and service levels.

A much more detailed breakdown of traveller types was possible than in the GLTS model, namely by five purposes (work, education, shopping, other and pensioner), on seven modes (car, bus, BR rail, LT rail, walk, two-wheels and taxi), and by three levels of household car ownership. Work and education journeys are taken to be in the peak-hours; all other purposes are taken to be in the off-peak.

The road speeds are given by speed/flow relationships established for each zone for peak and off-peak, and the public transport systems have a financial constraint to ensure that the revenue from the fares offered and service levels provided match the operating costs plus any subsidy.

The interesting point about the calibration of the model is that the household size, car ownership and trip generation rates are all estimated using time trends, using both GLTS surveys or national data where appropriate. This means that no verification procedure can be used since all the survey data is used in calibration.

In particular, it was noted that there is a difference in the procedures for surveying walk trips between the two surveys, which may therefore affect the calibration.

Perhaps more seriously, using a time trend to calibrate the change in trip rates between the two surveys, when there was a particular change in trip costs in the period, may well bias the forecast estimates of trip rates, when there will be different changes in trip costs. In particular, there is no mechanism in LAM for changing trip rates if trip costs change. The trip rates are, however, not being linearly changed with time, but logistically (i.e. changed to an asymptotic value); this restrains them from producing implausibly large changes. The description of the LAM concludes that this is an area which requires further attention.

The model is a standard transport model, with a power form in the distribution and modal split functions, following the precedent in Oldfield, Bly and Webster (1980). This used the ratio of generalized costs rather than the absolute differences as in the more usual exponential form because this was thought to be more useful in the range of problems to be studied.

Only the work trips are doubly constrained; that is, the origins

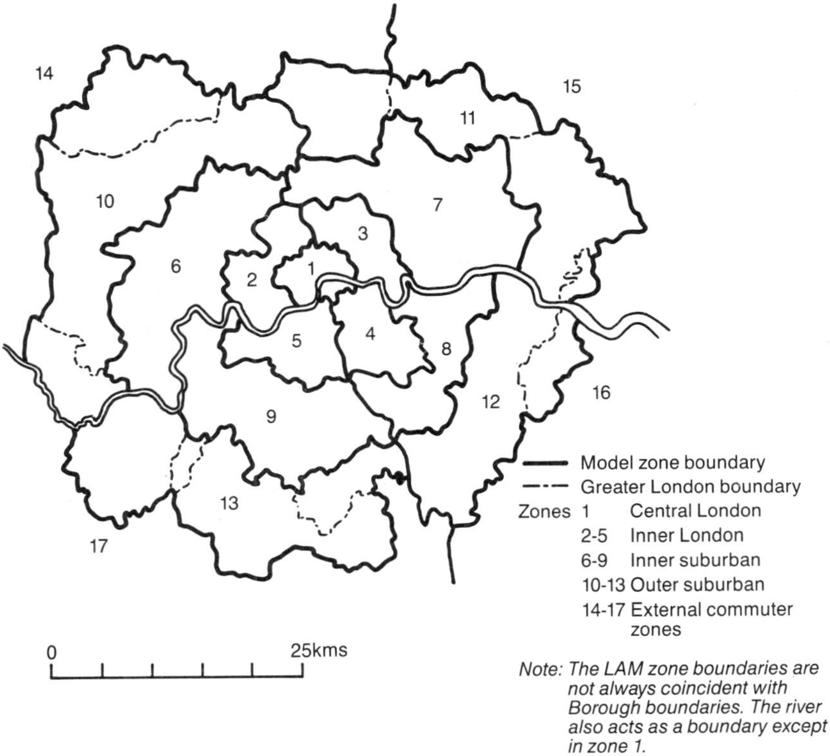

Figure 8.1 The London Area Model zoning system

and destinations in each zone are both fixed. All other trip purposes have a singly-constrained form (i.e. only the origins are fixed) and the elasticities of demand with respect to generalized cost determine how and how far the trips go. The initial pattern is calibrated on a zone-to-zone basis with a set of calibration factors; this is necessary because the generalized costs do not fully reflect the availability of the modes between the zones because of the coarse size of the zones.

The public transport generalized costs reflected the investigation reported in Bly and Oldfield (1986a); that is, optimizing services and fares within the financial constraint, and including a bus capacity limitation effect. The mean bus speed was, however, related to the mean road speed with an allowance for time spent at stops. There was, therefore, no inbuilt effect of a bus-lane programme (i.e. to speed up the buses), but this could have been built in without too much difficulty.

Initial calculations using the model, Bly and Oldfield (1986b), showed that:

> a policy of halving public transport fares gave a net benefit of 17p per additional pound spent on subsidy, patronage on public transport increased by 16 percent overall and car use decreased by four percent overall. In the central area, car use dropped by 13 percent, leading to increased road speeds of eight and three percent, peak and off-peak.
>
> a policy of charging £5 per day entry fee to the central area, or £3 for central area residents every time they make a two-way journey across the central area boundary, would reduce car use in the central area by 27 percent, with a corresponding increase in road speed of 14 and 8 percent, peak and off-peak. There is a net benefit equivalent to 15p per £1 of charge levied, but this conceals losses of benefit to car users greater than the charge imposed.

When the London Planning Advisory Committee (LPAC) came to consider how they should fulfil their responsibilities to offer advice to the Secretary of State for the Environment on transport in London by mid-1988, it was completely understandable that they should turn to LAM, as this was precisely the function for which the model had been set up by the TRRL, as we saw from the brief.

LPAC therefore commissioned the MVA Consultancy (MVA), in conjunction with Colin Buchanan and Partners (CBP), to use the LAM to provide an evaluation of a specified set of scenarios. The TRRL provided the model but took no part either in specifying the scenarios or in their evaluation. The report produced (MVA *et al*, 1988) is called *Transportation Strategic Advice: Scenario Testing Exercise* (TASTE). Essentially, with a time horizon of 2001, only policies which could be implemented by that date were chosen. Four basic scenarios were adopted, emphasising different aspects of choice, with a number of variants.

(1) This was a package of currently advocated policies, without ensuring that they formed a cohesive whole or that they represented a considered approach to achieving a given set of objectives, although no obvious inconsistencies or contradictions were noted.

The package consisted of capital investment (both road and rail) and traffic management schemes which were likely to be implemented by 2001, together with the continuation of current policies on subsidies to public transport operators and the likely consequences for fares and service levels, and the replacement of existing assets where appropriate to maintain those service levels. The elimination of company car subsidies was assumed, as was the deregulation of buses in London, with a particular consequent impact on costs.

This base scenario was assumed to cost £3.4 billion in 1986 prices, of which £1 billion is road investment.

(2) This was a package based on public transport infrastructure improvements, on maintaining fares at 1986 levels and decreasing BR off-peak service headways to 15 minutes at most, with improved on-street parking enforcement in central and inner London.

Some committed road and rail capital investments were scrapped; a package of rail investments thought to be more in keeping with LPAC strategic objectives of improving accessibility was chosen.

The net increase in capital cost of the schemes over the base was £1.3 billion.

(3) This was a package based on road infrastructure improvements (particularly in outer London) together with computer-controlled traffic management in inner and central London. Improved parking control on-street was to be balanced by more off-street parking.

The net increase in capital cost of the schemes over the base was £1.5 billion.

(4) This package was based on road pricing in the central area, together with traffic management in central and inner London as per scenarios 2 and 3 but without increasing off-street car parking, and a peak surcharge on public transport central-bound journeys.

There is therefore no additional capital investment in this scenario over the base.

The variants over the various packages involve the following:

(2a) A road pricing option for the central area.
(2b) Halving the subsidy required for scenario 2.
(3a) No road capacity increase from traffic management (but instead environmental improvements).
(3b) A road pricing option for the central area.
(4a) A road pricing option for the inner area.
(4b) Fares at 1986 prices (except for central surcharge) and train service increased by 20 percent.

In order to test these scenarios, the consultants made a number of changes to the LAM, by changing some of the assumptions underlying the in-built relationships, by postulating additional mechanisms and by testing the sensitivity of the results to the assumptions. This is exactly what the brief for the model given earlier specified as the necessary requirements for the model to possess.

The LAM was calibrated on 1986 data and forecasts for 2001 made on the basis of a number of disparate sources. Some of these forecasts were used to control the LAM totals, rather than using internal relationships established in the LAM. Essentially, a rise in London population of 0.4 percent and employment of 5.3 percent was forecast, rather than the decline that had been the case in the forecasts made in the early 1980s. Cars were forecast to increase by 27.8 percent in the 15 years and unemployment to remain at 10.0 percent.

Just as with CRISTAL and Lynam and Everall's (1970a,b) attempts to reconcile traffic flows on the schematic network with the passenger distance actually travelled, so too in LAM there is a problem in reconciliation. This is partly due to a main-mode assumption (i.e. the whole length of the journey is assumed to be on the mode which forms the longest part of the journey), and partly due to the ignoring of visitors and of people not living in households.

Some indication of the importance of these effects can be seen in the factors applied to the modal distances from the household survey to reconcile them with the measured distances actually travelled on the networks.

Table 8.1 Reconciliation factors for the main-modes in LAM

Car	1.431
Bus	1.596
BR rail	1.603
LUL rail	1.234
Taxi	1.254
2-wheel	1.06
Walk	0.98

MVA et al (1988, Paragraph A2.22)

The proportions of journeys by purpose occurring in the peak and off-peak were altered to conform to the observed proportions in GLTS 1981, rather than for work and education journeys to be in the peak and all other off-peak as in the original LAM. The generalized costs for each purpose are therefore a weighted average of peak and off-peak costs.

The speed/flow relationships in LAM is based on work by Duncan *et al* (1980) nationally and Fairhurst and Mack (1981) in the London area. It implies a fall in speed of 1.55 km/h for each 10 percent increase in flow from the base observed flow in the peak and proportionately less in the off-peak according to the ratio of traffic flows. The consultants initially modified this relationship to make the relationship more sensitive to increases in flow as flow increased above a critical level. The method was simply to increase the fall in speed by a factor of 5 once the speed had fallen below critical speeds defined for each ring. These were 20.5 km/h in the centre and inner ring; 25 km/h in the inner suburbs and 30 km/h in the outer suburbs. Later sensitivity testing examined the criticality of this assumption, as discussed below.

As LAM is a non-network model, new road capacity has been incorporated by changing the speed/flow relation in proportion to the increase in capacity provided in each zone. This is mathematically straightforward, but liable to considerable error. New rail capacity is taken to modify the zone-to-zone in-vehicle times by a subjective assessment. Bus deregulation was assumed to cut unit costs by 15 percent for big buses; minibuses, one third of the total, are assumed to cost 2/3 of big buses per vehicle-km; a 20 percent higher level of service is run for the same demand level.

The company car subsidy could only be dealt with by a modification to the central area parking sub-model, to reduce the amount of free parking by half.

The consultants also found that through car traffic in the central area was only 10 percent in LAM as compared with about 30 percent in reality. This is, as will be recognized, the old problem of assessing the true costs of central area car trips.

The evaluation was done by reference to scenario 1 as a base. Table 8.2 shows the results for the various scenarios. Department of Transport assumptions were used throughout to set the values to be attached to changes in price levels, values of time, vehicle operating costs and so on, taking the mid-points of ranges where appropriate; for example for economic growth, fuel prices.

A term is added for accident costs, based on vehicle-km and adjusting for road type. Sixty percent of this is added to the costs of travel and 40 percent to the public costs.

Whereas petrol prices increase by 30 percent in all scenarios, public transport fares rise according to the following Table 8.3., either as specified in the input or as a result of the equilibrium established in LAM.

It can be seen that the base case (scenario 1) performs less well than any of the others. The rail investment (scenario 2) shows a high level of benefit, but at a high cost, most of which is in keeping fares constant. The road investment (scenario 3) shows a low level of benefit, and a low level of cost. Much of the benefit (some £40 million) actually comes from park-and-ride facilities on the M25 shifting car users to rail. Without this element, the benefit/cost ratio would be 1.1.

The road pricing scenario (4) shows that the financial gain to the public sector is nearly £300M pa. The yield of the road charge is £340 million pa. Car users save £302 million in time and £83 million in operating costs but incur £461 million in charges, a net loss of £76 million, together with a saving in accident costs of some £7 million. The central area peak surcharge on public transport fares, however, permits fares to be kept low compared with scenario 1. The most significant benefits go to businesses and freight travel, which benefit to the amount of £82M pa, and to the poorer households in London, who benefit by the faster bus services.

Table 8.2 Summary of scenario appraisals in TASTE

£M pa resources at 1987 prices in 2001	*2*	*2a*	*2b*	*3*	*3a*	*3b*	*4*	*4a*	*4b*
Travellers' time etc									
Car	177	343	94	103	80	378	302	392	342
Bus	19	48	–29	15	11	69	63	90	33
BR rail	81	51	62	40	41	65	25	45	87
LT rail	118	99	114	0	0	0	7	12	129
Freight	4	10	3	2	2	11	7	12	10
Change mode etc	83	–20	45	6	5	–49	–41	–165	–19
Subtotal	487	539	291	165	136	480	371	395	599
Fares, taxes and charges									
Car	5	–463	6	9	9	–460	–461	–1168	–463
Bus	96	95	49	5	3	23	19	23	70
BR rail	270	271	124	1	2	37	37	56	107
LT rail	168	168	72	–5	–3	2	14	26	37
Change mode etc	–76	78	–50	–1	–2	75	60	244	51
Subtotal	462	148	201	9	9	–323	–331	–820	–199
Vehicle operating costs (less tax)									
Car	59	117	28	8	6	84	83	147	–101
Accidents (60%)	9	19	6	3	3	12	7	18	22
Total benefits	1015	820	525	184	153	248	127	–264	520
PT fares—costs	–679	–621	–281	0	0	6	5	6	–207
Taxes, fees	51	300	18	–6	–8	282	286	720	286
Infrastructure	–93	–93	–93	–126	–126	–126	0	0	0
Accidents (40%)	6	13	4	2	2	8	5	12	15
Total costs	–715	–401	–352	–130	–132	170	295	737	94
Net social benefit									
B+C	300	419	173	54	21	78	422	473	614
B+1.2C	157	339	102	28	–5	452	481	621	633
B+1.5C	–57	218	–3	11	–44	503	570	842	661

The columns will not sum due to rounding

Rows with items less than £10M have been omitted

Infrastructure costs have been annualised over 40 years at 7%pa.

Changed mode etc includes the travel costs of those who change mode, destination etc compared with scenario 1.

Net social benefit is expressed as total benefits plus a ratio of the total costs to express the fact that benefits must exceed costs by a certain ratio to be worth doing.

Table 8.3 Fare changes in the scenarios in TASTE

1986=100	*1*	*2*	*2a*	*2b*	*3*	*3a*	*3b*	*4*	*4a*	*4b*
Bus	185	100	100	141	181	182	164	146	142	100
BR rail	161	100	100	134	162	162	154	116	112	100
LT rail	165	100	100	138	167	167	165	109	104	100

Table 8.4 Percentage changes in passenger-km in TASTE (1986=100)

	1	*2*	*2a*	*2b*	*3*	*3a*	*3b*	*4*	*4a*	*4b*
car	22.3	17.0	11.6	20.1	22.8	21.6	17.2	16.1	8.7	14.1
bus	-31.3	-15.9	-11.2	-28.8	-29.8	-30.3	-21.0	-21.3	-13.8	-16.5
BR rail	-5.5	4.9	12.3	-0.7	-5.3	-5.1	3.1	3.3	8.5	6.5
LT rail	-4.8	5.2	7.9	1.4	-6.4	-5.9	-4.5	0.1	4.9	4.6

The passenger-km by the various main modes change in the various scenarios as given in Table 8.4.

In the base scenario, a fall in bus passenger-km of over 30 percent is estimated, and almost as much in scenario 3, the road investment scenario; in contrast, in scenario 2, the rail investment and constant fares scenario, bus passenger-km falls only by 16 percent. Likewise, rail falls by about 5 percent in both 1 and 3, but rises by about 5 percent in 2. It should be remembered, however, that LAM has no mechanism for changing trip rates according to trip costs, so the amount of public transport travel may well be underestimated in scenario 2 (and 4b) in particular. In scenario 4, the road-pricing scenario, bus passenger-km falls by 21 percent and rail rises slightly. Car passenger-km rises by almost as much in each scenario, 22 percent in 1 and 3, 17 percent in 2 but only 16 percent in 4. Interestingly, therefore, the massive investments in roads and traffic management infrastructure in 3 has hardly affected the rise in car use, but has slightly reduced the decline in bus use, but worsened the decline of LT rail use relative to scenario 1.

Scenarios 2a and 3b (the road pricing variants) show reductions in car use and increases in public transport use as we might expect. Scenario 2b, halving the public transport subsidy, shows a similar car and bus use to scenario 1 but a rise in rail use. Scenario 3a shows very little difference to scenario 3, except that benefits drop substantially.

Scenario 4a, road pricing in inner London, shows car use dropping the most of all scenarios, and bus least, with fairly high rail use. The revenues from the road tax substantially outweigh the loss of benefit to car users. Scenario 4b, with no increase in fares, and improved rail services, shows a net benefit of over £500M achieved with a net financial gain of nearly £100M.

It is of particular interest that the difference with scenario 1 shows that an extra £393 million of benefits is achieved at a net financial cost of £201 million, a benefit/cost ratio of nearly 2.0. This implies that a better design of the road pricing/public transport subsidy scenario could produce an even higher benefit/cost ratio. In the TASTE report (MVA, 1988 *et al*, pp.75–76) it is suggested that:

> 'a parallel treatment with more subsidies for buses as well as rail modes could further enhance this scenario, possibly by a very significant amount, especially as with less road congestion buses would become much closer substitutes to rail for many short/middle distance trips. A greater emphasis on buses would also tend to give a higher proportion of the benefits to lower income households.'

Thus in the TASTE report we see the first full modelling of a road pricing/public transport subsidy combination giving a high benefit/cost ratio, even though the LAM techniques have some severe short-comings, chief of which is the lack of any feed-back between trip costs and trip generation rates. This may well tend to under-estimate the benefit from this policy combination.

In a further report to LPAC, MVA & CBP (1989), the sensitivity of the LAM techniques to some of the assumptions was explored, together with some further studies of some of the major options outlined in the TASTE scenarios. By this stage, the model was updated by TRRL, mainly to change the definition of work trips.

The parameters in the speed/flow curves, the level of rail costs and the level of bus costs through deregulation were explored and were not found to alter the main conclusions about the relative merits of the scenarios. The level of road cordon charges and public transport surcharges to enter the central area were varied over a wide range and showed that, whilst the users' benefit peaked at a cordon charge of around £3 in 1986 prices, the net social benefit continued to rise well beyond that value—without using the revenue to subsidize public transport. Varying the level of fares and services showed that the benefits could be increased for a given level of subsidy more by increasing services than by reducing fares, which is what one would expect from the TRRL (1980) study.

A study of the various systems of road pricing concluded that electronic road pricing, as suggested for Hong Kong (Dawson 1983), but modified to overcome the privacy problem by using pre-paid charge-cards, would be likely to be the most effective system, using a number of cells to cover central and inner London.

A study of the various options for rail investment outlined the principles which the consultants thought should underlie rail investment. These included:

> 'that it is more important to provide significant new or better connections than to duplicate existing ones. This includes faster cross-London links.
>
> 'that schemes designed only to relieve overcrowding are unlikely to be the best.
>
> 'that schemes must build on major opportunities such as the need to serve Heathrow and the Channel Tunnel terminal.
>
> 'that it is sensible to try to open up areas poorly served by rail.'

They suggested a number of new LT tube and BR cross-rail schemes. Seven possible strategies were devised, ranging in cost from £3 billion to £1.5 billion.

More interestingly, an initial network of some 50 route miles of tramways along some of the central London streets and on the major radials in inner London—costing some £300M—was suggested as a replacement for some high-density bus corridors, with considerable benefits to users as compared to buses, together with major environmental benefits.

A study of the effects of public transport subsidy on road congestion using further test runs of the LAM found that the cross-elasticity of road traffic levels to the levels of fares was about 0.1 in central/inner London, somewhat higher than Lewis's (1977) value for central London. The consultants did not feel, however, that much confidence could be put on this result given the limitations of the LAM.

In parallel to these LPAC studies, as part of their own strategic responsibilities after the abolition of the GLC, the Department of Transport also set up four transport assessment studies in 1984 of particular parts of London covering a number of boroughs where traffic congestion was thought to be particularly bad, and where better road through routes were particularly desired. Three of these areas are in inner London, east, west and south of the centre; the fourth is in outer London covering a corridor in south London. These assessment studies were done by different consultants. Phase One of the Process was defining the problems, with surveys and so on, and reported in November 1986. In particular, these studies showed that new roads, by themselves, could not solve the problems.

The GLTS model, now renamed the LTS model, was extensively used in all these transport assessment studies to provide a strategic framework for the more detailed traffic models of the local areas. Phase Two A of the Process was suggesting alternative options and reported in June 1988; Phase Two B was suggesting solutions and was reported in December 1989. Subsequently, a number of other assessment studies have been begun for other areas where traffic congestion is thought to be particularly bad.

Again in parallel, with the recovery of central London commuter traffic after 1982 (Evans and Crampton (1989)), although not readily apparent by 1985, the issue of overcrowding on the rail system became highly critical; eventually the Department of Transport, in conjunction with the operators, reacted by setting up a further study, as indeed the GLC 10-year Plan had recommended. I will describe this rail study first.

The Central London Rail Study (1989) was in fact not commissioned until 1988, and is a joint study by the Department of Transport, British Rail Network SouthEast (NSE), London Regional Transport and its subsidiary—London Underground

Limited (LUL). The consultants, using the LTS model, tested a number of alternative schemes to see which relieved congestion on the existing railways most effectively, including several versions of the Cross-Rail scheme, connecting NSE suburban services, and new cross-centre LUL tube lines.

It should be remembered that these schemes were designed to relieve congestion on the railways; they are not necessarily the best schemes that could be designed to improve transport in London, according to the principles outlined above in the LPAC study.

With a 2001 planning horizon, based on 1987 estimates, population was forecast to remain stable, though with a growth in Docklands and a corresponding fall elsewhere. Employment, however, was estimated to grow by between 100,000 and 150,000 in central London, by between 120,000 and 150,000 in Docklands and by between 60,000 and 120,000 in the rest of London. Only the lower forecasts were taken for the purpose of estimating traffic flows.

This is a very conservative estimate, given the rate at which employment has been increasing since 1982, and given the structural analysis of Evans and Crampton (1989). But it is a distinct change from all but the very first forecasts in the LTS of 1968—20 years earlier—before the decline in central employment from 1962 to 1982 was realized. Car ownership was forecast to increase by 20 percent by 2001, and real fares to move in line with GDP, so that a growth in London-wide car traffic of 10 percent in the morning peak is forecast by 2001.

A substantial programme of road construction was assumed, including the upgrading of the North Circular Road, the construction of the East London River Crossing and a new road network in Docklands. Extensions of the Docklands Light Railway to Bank, Beckton and Lewisham are assumed, with increased capacities and frequencies.

A particular feature of the modelling was that, for the first time, a notional cost of congestion was included on the rail system where appropriate to reflect the short-run increase in costs with increasing flows of passengers.

The modelling showed conclusively that the cost-benefit ratios for a number of cross-central London rail schemes were greater than unity. A list is given below in Table 8.5. The lines are shown in Figure 8.2.

Table 8.5 Benefit-cost ratios for cross-rail schemes (1989)

A Costs (£M 1988 prices)

(i) capital cost of construction
(ii) rolling stock costs
(iii) equivalent annual capital cost
(iv) annual operating cost
(v) total equivalent annual cost

Notes

1. capital cost discounted at 7 percent pa over 60 years (35 years for trains)
2. construction cost spread over a 5 year period prior to opening

Single line schemes	*(i)*	*(ii)*	*(iii)*	*(iv)*	*(v)*
Chelsea–Hackney	1000	330	110	40	150
Jubilee extn to Ilford	560	80	55	15	70
Thameslink Metro	260	70	25	25	50
East–West Crossrail	870	15	75	10	85
North–South Crossrail	840	55	75	10	85
City Crossrail	650	50	60	10	70
Two Line packages					
Full Cross (E–W + N–S)	1710	70	150	20	170
E–W Cross + C–H	1870	345	185	50	235

B Benefits (£m. 1988 prices)

(i) time savings
(ii) crowding relief
(iii) total public transport users' benefit
(iv) road users' benefit (road congestion and road accidents)
(v) generated rail revenue (new journeys)
(vi) total benefit

Single line schemes	*(i)*	*(ii)*	*(iii)*	*(iv)*	*(v)*
Chelsea–Hackney	60	30	90	20	130
Jubilee extn to Ilford	5	25	30	5	40
Thameslink Metro	55	20	75	15	105
East–West Crossrail	30	70	100	20	140
North–South Crossrail	40	15	55	10	75
City Crossrail	55	40	95	20	135
Two Line packages					
Full Cross (E–W + N–S)	70	85	155	30	215
E–W Cross + C-H	90	100	190	40	270

continued

Table 8.5—continued

C Benefit–cost ratios

(i) *to public transport users*
(ii) *to road users*
(iii) *revenue gain*
(iv) *total*

Single line schemes	*(i)*	*(ii)*	*(iii)*	*(iv)*
Chelsea–Hackney	0.6	0.1	0.1	0.9
Jubilee extn to Ilford	0.4	0.1	0.1	0.6
Thameslink Metro	1.5	0.3	0.3	2.1
East–West Crossrail	1.2	0.2	0.2	1.6
North–South Crossrail	0.6	0.1	0.1	0.9
City Crossrail	1.4	0.3	0.3	1.9
Two Line packages				
Full Cross (E–W + N–S)	0.9	0.2	0.2	1.3
E–W Cross + C–H	0.8	0.2	0.2	1.1

Rows may not sum due to rounding

The assumption of constant trip ends in a doubly constrained trip distribution and modal split model leads to a peculiar quirk in the modelling, not apparent in the report but only in the technical documents. The rail improvements to central London cause an increase in road traffic on the North Circular Road and other orbital routes.

This is because there is not only a switch from road to rail in central London-bound traffic along the lines of the new or improved routes, thus relieving the roads in those areas; but also a diversion of some rail and road traffic to other destinations caused by the increased competition for central London workplaces from rail users on the new routes. The extra road traffic causes increases in traffic levels on orbital routes, and thus increases travel times on them. This is taken to offset the benefits from the reduction in traffic levels along the lines of the new or improved routes.

The mechanism of this effect has been explored by Beardwood (1989) using the East London River Crossing as an example. In this case, by contrast, with increased road capacity the new road users diverted onto the the new crossing and on into central London from

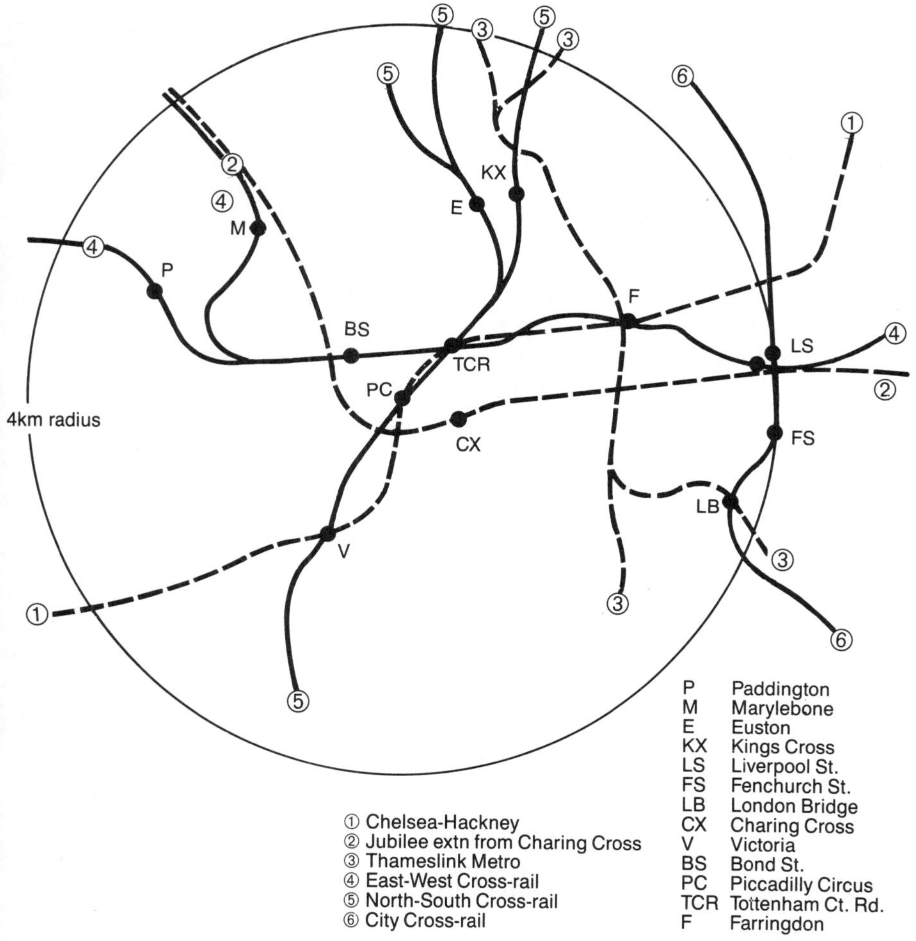

Figure 8.2 The Cross-rail schemes

rail cause a diversion of previous central London-bound road users onto orbital routes in just the same way. Only this time, the extra distance travelled because of the increased competition for central London workplaces is classed as a benefit. This inconsistency remains to be clarified.

The Conservative government refuses at present to consider cost-benefit appraisal for taking decisions on railway investment, although it does so for road investment, as in the East London River Crossing just mentioned. Its policy was laid out in the Department of Transport (1988b) paper Cm. 420. It stated:

> 'The Government expects to approve investment in railways, where the investment will earn a commercial return . . . In principle, the schemes able to earn the best returns on the capital invested should also be those which best meet the needs of passengers as measured by their willingness to pay for them.
>
> 'Some of the schemes under consideration, or parts of them, would also enhance land values or otherwise generate significant benefits for particular businesses or landowners. Such beneficiaries may wish to enter into negotiation with the railway operators with a view to influencing the choice of routes, or their alignments, or the siting of stations in return for a significant financial contribution to reflect the benefits they would derive . . .
>
> 'If the revenues which can be expected from fares, together with financial contributions from developers, are not sufficient to pay for a scheme, the Government will consider the case for a grant, if there are sufficient non-user benefits (eg in the relief of road congestion) to justify the amount requested to make the investment financially viable in accordance with the usual test discount rate.'
>
> Central London Rail Study (1989, p.22)

Relief from road congestion in central London, where road traffic is no more than 25 percent of the total daily movement, can, of course, be expected to be about 25 percent of the total benefit if there is the equilibrium in journey costs I have been discussing. The paradox in this position is that the more road traffic there is in the centre, and the less public transport, the more the relief from road congestion and the more support the public transport system would get from grant.

The impact of the suggestion that developers might wish to contribute financial support in order to influence line development can be seen in the East London Rail Study (1989), where the developers of the Canary Wharf project in Docklands proposed to contribute finance to an extension of the Jubilee Line, provided it went via Waterloo (the new Channel Tunnel terminal) and on to Stratford. This proposal is now being examined by Parliament, in preference to either of the Cross-Rail packages, and involves leaving an existing one-kilometre section of the Jubilee Line from Green Park to Charing Cross as a spur to the new main line, and abandoning the further tunnels towards the east already constructed or safeguarded, including tunnels at St Paul's and Fenchurch Street.

This is despite the fact that in the Central London Rail Study, the variant of the Jubilee Line extension studied there had a benefit/cost ratio of 0.6; that is, annual costs of £70 million and benefits of £40M. It would thus be a net loss to the community on the assumptions made and in the Central London Study it was proposed to discard this option! In the East London Rail Study, however, with a different alignment and with other assumptions, particularly with new employment forecasts in Docklands, the line had a benefit/cost ratio just over 1.0.

It can be seen in Table 8.5 that several schemes show benefit-cost ratios greater than unity; that is, they appear to be worth doing, even on the basis of benefits to existing rail passengers alone. Chief among these is the Thameslink Metro scheme, which has a benefit/cost ratio of over 2. This is, in fact, already going ahead in part, with a new station to replace Holborn Viaduct, the first central London terminus to be closed on 26 January 1990. This new station—St Paul's Thameslink—is being built underground with a possible future connection to the Central Line St Paul's station. A new property development will pay for the work. New connections at Blackfriars to the Waterloo and City Line and at Southwark to the new Jubilee line may be made, together with extensions to the track to London Bridge to enable a peak-hour service to be run. This requires Parliamentary approval which may be sought in the near future.

If, however, rail users were to be charged fares to cover the benefit from these new rail facilities, it would drive some potential users off the rail system and only a substantial fraction could be recovered, not all, as was shown by Dupuit's footbridge example in 1844.

The major effect of the rail schemes is not in reducing road or rail congestion but in increasing accessibility, that is in allowing people to reach more places within a given time or cost. This is in fact the major justification used in Paris for the Regional Express Rail (RER) system, a similar system in concept to Cross-Rail. It is of particular use in allowing people in East London to reach the West End, and in West London to reach the East End: something which at present is extremely difficult because of the slow interchange at the rail terminals.

At the end of Phase Two B of the four transport assessment studies for the four areas of London, and the Central London Rail Study, the LTS model was used in a final summary run of a maximum road option scheme as compared with a maximum rail option scheme for the Department of Transport for the forecast year of 2001 (MVA, 1989). Similarly, the LTS model was used for a final summary run of the maximum road option scheme as compared to the LPAC strategic advice scheme for the same forecast year: 2001 (MVA, 1990).

The base is in all cases a do-minimum representing a continuation of present government policies up to the forecast year and includes road and rail schemes as already noted in discussing the TASTE scenarios. There are, however, a number of slight differences in assumptions between the Department of Transport and LPAC, although in such matters as income, population and employment growth the same assumptions are used.

In the Department of Transport's tests, the maximum roads option represented the various highway proposals made by the four consultants in stage Two A of the transport assessment studies and are thus concentrated on the four areas, rather than being the best schemes that could be devised. The maximum rail option was defined before the Central London Rail Study reported, and included: a Cross-rail between London Bridge and Paddington and between Victoria and Kings Cross, the Chelsea-Hackney tube line, the Thameslink Metro, extensions and interconnections of BR and LUL lines (especially around the centre as new orbitals), and an LRT or tram system in Croydon.

As yet, the evaluation of these results has not been reported, only the traffic effects. Essentially, the modelling showed only limited effects on road traffic as a whole of the order of a few percent, for both options, which suggests that an infrastructure-only approach would not have a major effect on road congestion. There are, however, substantial shifts in traffic flows on the rail and bus networks as a result of the rail investment, as one might expect, with substantial changes in accessibility. Of particular interest is the manner in which the orbital railway improvements reduce the rail flows into the centre.

In the LPAC tests, the maximum road option includes—besides a closely comparable set of roads to the Department's option—various schemes for public transport improvement in the four areas such as: the Croydon LRT system, three new stations on the Thameslink service and slight improvements to the South and West London orbital rail services. The strategic advice option is based on an electronic road pricing scheme, comprising six cells (2–3 km diameter) for central London, an inner (5 km radius) cordon and an outer (8 km radius) cordon, with a charge of 50p to cross any cell boundary or cordon. Extensive rail investments comparable to the maximum rail option are also made in the strategic advice option, but with only one Paddington-Liverpool Street Cross-rail rather than two. Bus fares are held constant in real terms, whilst rail fares are assumed to rise by 30 percent, compared to income growth of 47 percent. A number of other more minor changes are made.

The LTS model is a peak-hour only model; in practice, the road pricing system would be all-day in the centre and in the peak direction on the two cordons. The 50p charge per crossing is designed to be equivalent to the £3 per day charge earlier shown to be optimal for maximizing users' benefits.

The results show that the maximum roads option has a negative value to the community (i.e. it costs more than the benefits) whereas the strategic advice option has a large positive value. In particular, the latter shows large increases in accessibility; in the outer suburbs to the south-east, north-east and west, accessibility to jobs rises by up to 35 percent, with other areas showing rises of between five and 15 percent, as we might expect from Cross-rail.

The LTS network model thus confirms what the LAM non-network model was demonstrating, that a policy of road pricing in the congested heart of the city coupled with subsidy to public transport is the most advantageous policy in terms of maximizing the net social benefit. Whilst these scenarios are not chosen so that the road price and the public transport subsidy conform to the notion that the user should pay the marginal price in order to maximize net social benefit; they demonstrate that there is indeed a road price and a public transport subsidy level which in combination would do just that.

It should be noted that one has to be extremely careful when evaluating public transport options where there is a decreasing cost with increasing flow. This does not conform to Beckman *et al*'s criteria, and may result in a set of possible equilibria rather than a unique equilibrium. Obviously one requires the lowest cost equilibrium point.

I argue therefore that all investments in transport infrastructure should be made on the basis of a cost/benefit analysis, as we have seen for example in the Central London Rail Study. The policy to be applied to the price charged for use—once the facility is built—should, however, be on the basis of charging the marginal price, not an average price. In public transport infrastructure, attempts to recover the cost of construction as well as operating costs will result in the facility being underused, thus losing benefit to the community. If the facility has a price charged which is lower than the marginal charge (which is the case on congested roads), then the facility will be overused, thus also losing benefit.

This was exactly Hotelling's (1939) point about financing investments in transport infrastructure out of tolls, developing the point made by Dupuit (1844 et seq), whom he quoted from the collection with comments by di Bernardi and Einaudi (1932). Hotelling, in his presidential address to the meeting of the Econometric Society in Atlantic City in 1937, stated:

> 'The actual calculation of (the total net benefit) (in a transport investment) would be a matter of estimation of vehicular and pedestrian traffic originating and terminating in particular zones, with a comparison of distances by alternative routes in each case, and an evaluation of the savings in each class of movement.
>
> 'Determination whether to build (the transport infrastructure) by calculation merely of the revenue obtainable from tolls is always too conservative a criterion. Such public works will frequently be of great social value even though there is no possible system of charging for their services that will meet the cost.'
>
> Hotelling (1938, pp.247-48)

Hotelling went on to argue that investments should therefore be made on the basis of a cost/benefit analysis; if benefits exceeded costs, the investment should proceed, provided there were no other calls on public funds with better benefit/cost ratios. Once the investment had been made, however, the prices that should be charged should be the marginal prices—not the average prices—in order to ensure that the resources of the community were used in the most efficient manner. Taxes or subsidies should be used to ensure that the price charged was the marginal price whilst the operator received the average price.

Wheaton (1978) also argued that the wrong investment priorities as between the roads and mass-transit would be determined if the pricing system underpriced roads, thus leading to overinvestment in roads. He suggested that decision rules based on second-best criteria (i.e. subsidizing public transport) could overcome this bias. Friedlaender (1981) shows that these rules can become quite complex.

I argue, with Hotelling, that both taxes on roads and subsidy to public transport—to bring their prices to marginal prices—gives the most efficient use of resources.

9 *Transport Revolutions*

In Smeed's (1961) path-breaking paper, he begins by noting the contribution of Clark (1958) to the discussion of the influence of transport on the way that cities spread out through time. Clark (1951) had used a particular mathematical model, the negative exponential model of density with distance from the city centre, to estimate the rate at which cities spread out, and their central densities decline, with time. He used as a prime example the population statistics from the censuses in London, starting from 1801.

This was in part no more than a formal expression of the observation of Bleicher (1892) that city densities decline with distance from the centre, based upon his data from German and other European cities, and especially Frankfurt-am-Main, but going back to the Middle Ages. Bleicher's observation was that:

> 'Die rasche Abnahme der Dichtigkeit der Bevölkerung mit der Entfernung vom Mittlepunkte ist für unsere, der Zeit ihrer Entstehung nach auch sehr alte Stadt geradezu charakteristisch.'
>
> (The rapid decrease in population density with distance from the centre is very characteristic of an old city like ours.)
>
> Bleicher (1892, p.102)

Weber (1899) extended the analysis considerably, especially with US and other cities. Blumenfeld (1949), using Thompson's (1947) data, confirmed the pattern over very much larger metropolitan areas.

Finally Clark (1951), after a review of previous work (but not including Bleicher's) began his paper with:

> 'two generalisations the validity of which is now universally recognised.

(1) In every large city, excluding the central business zone, which has few resident inhabitants, we have districts of dense population in the interior, with density falling off progressively as we proceed to the outer suburbs.

(2) In most (but not all) cities, as time goes on, density tends to fall in the most populous inner suburbs, and to rise in the outer suburbs, and the whole city tends to "spread itself out".'

Clark then went on to define the negative exponential model. It is worth noting, however, that an ambiguous sentence by Kant (1962) led to some confusion *viz*:

> 'Clark kam zu demselben Ergebnis wie H. Bleicher (1892, S 98 ff) d.h. daß die Bevölkerungsdichte im allgemeinem mit der Entfernung vom Stadtmittelpunkte abnimmt und sich zudem durch folgende Exponentialgleichung ausdrücken läßt.'
>
> (Clark came to the same conclusion as Bleicher, namely, that population density falls with distance from the centre, and then went on to state the exponential model thus.)
>
> Kant (1962, p.346)

Unfortunately, Newling (1964) chose to interpret the sentence as a statement that Bleicher had also used the exponential model. This caused a chain of secondary attributions (including by Clark himself (1967, p.341)) until the problem was clarified by Edmonston (1975, p.6). There are, however, still tertiary attributions following from Clark's reference eg Ben-Akiva *et al* (1985).

The model is defined by the relation between d_0, the central density of the conurbation, and the rate of decline λ of density d with radius r. Thus

$$d = d_0 e^{-\lambda r}$$

Golob (1982) has given an extensive review of subsequent applications.

The gross population densities for London in 1971 and 1981 are given in Figure 9.1, using census ward data for the South-East

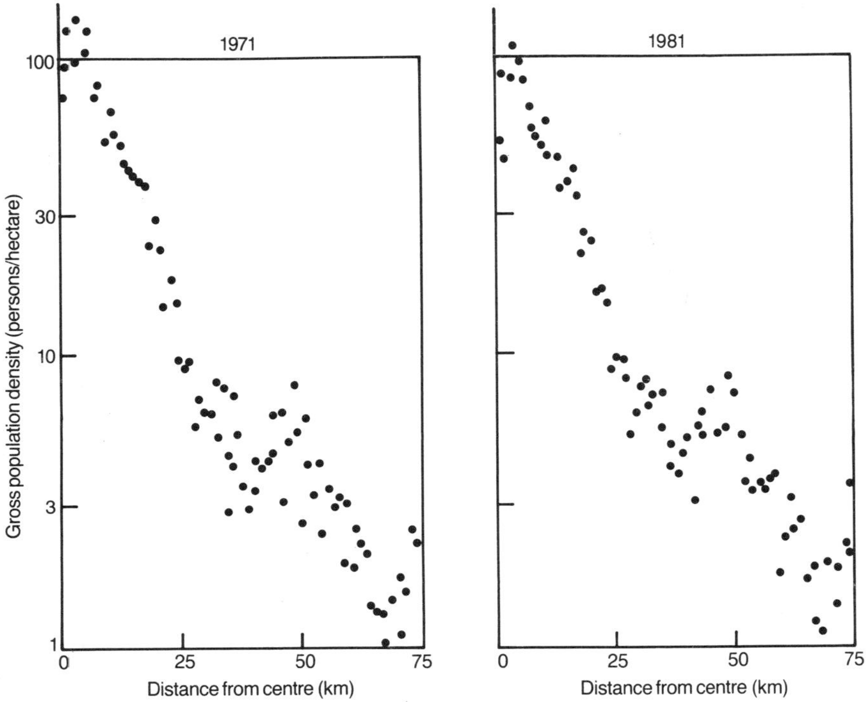

Figure 9.1 Gross population densities by distance from the centre of London 1971 & 1981

region (about 3,500 wards). Population is summed over annuli of 1 km width, being located by ward centroids.

The model does not, and cannot, fully represent the population distribution for a city, since no city is built on a perfectly featureless plain—indeed cannot be so built because the essence of location is the uniqueness of the site. This has led many workers to search for more complex mathematical descriptions of the spatial distribution.

In particular, the formation of the central services of the city has often led to a 'crater' effect for population at the centre—as noted already—with offices forcing the population away from the centre. (Initially, this 'crater' would have been formed by manufacturing and then wholesale and retail activities—increasingly now it is becoming service industries.) d_0, the central density in the model, is thus the hypothetical limit to which densities in the inner suburbs are tending.

At the other end, in the rural hinterland, a web of transport links will gradually enmesh formerly independent villages and towns. Hampstead and Clapham in London, now part of the inner city, were in the first census of 1801 separated from the City of London by some miles of open country, and an hour or so of travel. Even the name of one of the inner Boroughs, Tower Hamlets, recalls this pastoral heritage.

Each of these sites can act as nodes for growth of population, so that the population density beyond the continuous built-up core of the urban region does not fall off so fast as the negative exponential model would suggest.

Indeed, in the case of London, the fall beyond the built-up core appears to be modelled rather better by a power function model. The transition from the negative exponential model to this power function model occurs at the edge of the continuous built-up area. A number of authors have discussed this transition in model form, among them Bogue (1949), Weiss (1961), Ajo (1964, 1965), Newling (1966, 1969), Casetti (1969) and more recently Parr (1985).

The extent of the city is given by the radius at which it merges into the agricultural or even untilled land densities of its surroundings. Around London, where agricultural densities are $\sim$ 1 person/hectare, such densities are now hardly reached ($\sim$ 90 km) before the coast to the east and south or before the effects of the midlands conurbation are felt to the north-west.

For a strategic model of population density (such as the negative exponential model) to be useful in attempting to relate the impact of transport revolutions on the spreading out of the city, there must be some feature of the model that follows a time trend; a dynamic process. In the initial stages of industrialisation—as Clark (1951) noted—a rush from the land to the city causes central density to increase markedly as well as the spatial extent of the city. In the later stages, central density declines as the spatial extent increases.

Weiss (1961, equation 32) suggested that Clark's later data could be used to define a relation between d_0 and λ. Newling (1966, 1969) suggested other variations. I follow the extensive work from Bussière (1968) through to Bussière and Stovall (1981) where the central density is related to the rate of decline of density by

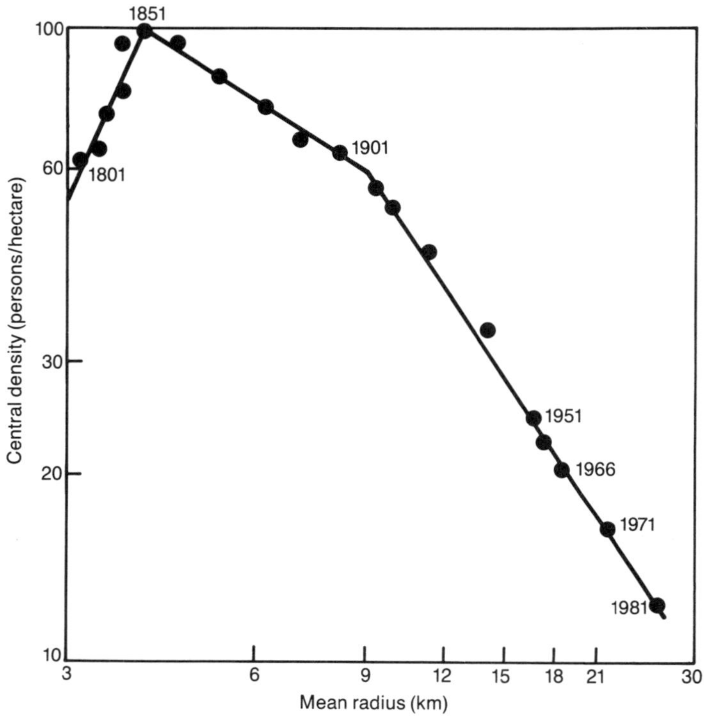

Figure 9.2 Relation between central density and mean radius of London 1801–1981

$$d_0 = a\lambda + b \qquad (a, b \text{ constants})$$

For London, Bussière and Stovall (1981, equation 8.5) give for 1901–71 the equation

$$d_0 = 318{,}939\lambda - 13{,}400$$

This has a maximum population to infinity of only 11,924,000 when the mean radius of the city $2/\lambda$ is 23.8 km. This is already the limit of the built-up area, beyond which as I have already noted the model is inapplicable.

We can see that there are two breaks in the relation, one at 1850 when the central density stops rising and one at 1900 when the slope of the relation increases. These two break-points must be explained by the transport technologies of the time.

It is interesting to extrapolate from the present position of the relationship for the next hundred years. The difficulty we face is to estimate whether the time trend established by Bussière is still relevant once the mean radius of the 'negative exponential' city has passed the edge of the built-up area for which the model itself is relevant. Indeed, Bussière and Stovall (1981, p.138) have already noted:

> 'Si la population de Londres continue à croître, ce sera nécessairement selon un mode différent.'
>
> (If the population of London continues to grow, it will necessarily be in a different manner.)

This different mode is, they suggest, according to a new slope of the linear relation connecting central density and the rate of decline of density.

I, on the other hand, suggested in Mogridge (1985) that a logarithmic form might be more appropriate in the future, thus

$$\log d_0 \ = \ a' \log \lambda + b'$$

I used an estimate of the rate of decline of central density of 25 percent per decade, and a rate of expansion of the city of 19 percent per decade (both calibrated from the 1971–81 data). The resulting population of the 'negative exponential' city thus increases at 6.5 percent per decade (1.19 * 1.19 * 0.75).

I forecast up to the year 2081, by which time the central density had fallen to 6.94 persons/hectare (from 124.00 in 1981) and the mean radius of the city had risen to 152 km from 26.1 km; consequently the total population had risen to 25.2 million from 13.3 million! It will no doubt be noted that a mean radius of this size puts a substantial fraction of the 'London' population on the continental side of the Channel in Northern France, Belgium and Holland.

Such a forecast is bound to be wrong because it makes no direct connection with the transport technologies available, but it does give an idea of what the current rate of expansion of the city into its surrounding countryside would be capable of achieving in 100 years if there were an appropriate technology.

Clark's (1951) observation—that density falls in the inner suburbs and rises in the outer, and (1958) that the turning point within which density falls and outside which it rises had been continuously increasing in distance from the centre of London since 1851—can be put into a more precise form using the dynamic equation.

In Clark (1967, p.357) these turning points are at the 'pivotal' density and radius. For the 1981–91 values given in Mogridge (1985), the pivotal density is at 20.73 persons/hectare and the pivotal radius at 23.38 km. (This is slightly less than the mean radius of the city: 28.43 km.) Thus the population density will continue to fall for all areas inside the former GLC boundary, apart from those where special factors apply such as in Docklands.

It is interesting to note that Craig's (1975) study of the 1931–1951–1961 censuses (using a sample of wards in all areas of Great

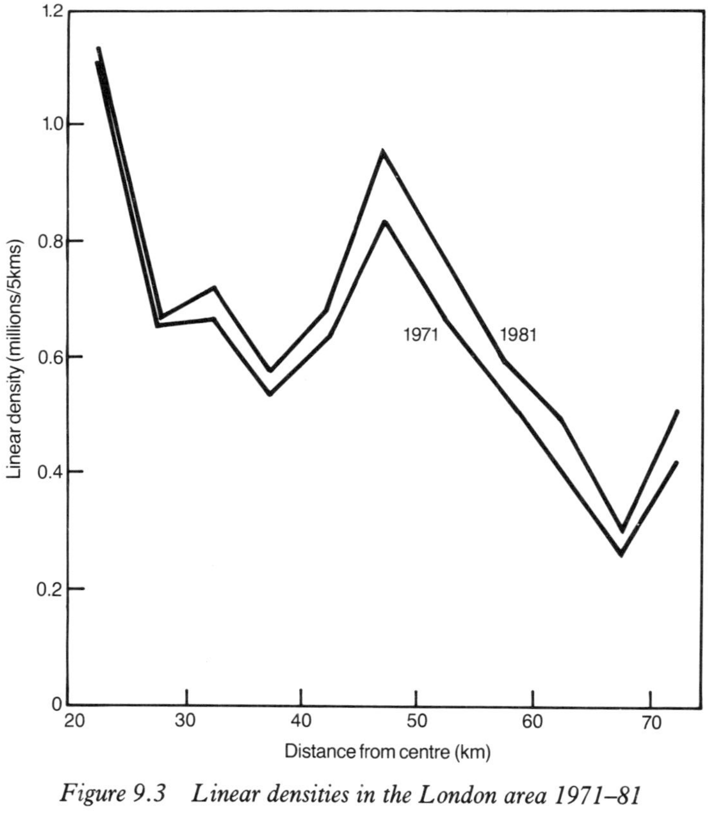

Figure 9.3 Linear densities in the London area 1971–81

Britain) is consistent with these values for pivotal density, as is his later study (1980) using the 1971 census. Holroyd (1984) has given estimates for the 1981 census. Thus pivotal densities in all GB cities seem to be at about the same level, irrespective of the size of city and thus of the pivotal radius.

If we examine in Figure 9.3 the linear density for London (the number of people contained within an annulus of unit width), we see that it is more precise to say that density first declines steeply with distance from the centre; beyond the 'green belt' there is a rise again in the 40–50 km ring, a further fall to the 70 km ring and then a further but smaller rise. We cannot go beyond this ring without going beyond the South-East region.

The influence of the 'green belt' therefore seems to be to depress densities immediately around the built-up area; the ring of 'first generation' new towns then raises linear densities again. Likewise,

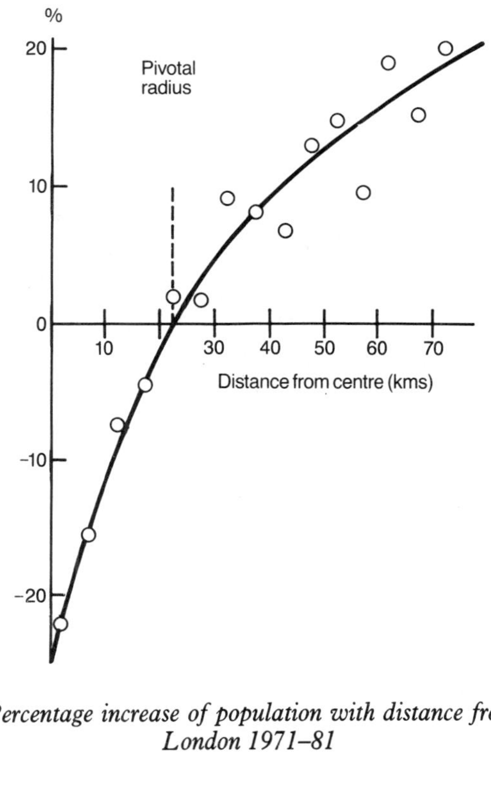

Figure 9.4 Percentage increase of population with distance from the centre of London 1971–81

the ring of 'second generation' new towns causes a further rise in the linear densities beyond 70 km.

The rate of change of density, however, continues to rise with distance from the centre right out to the edge of the region and beyond, as is shown directly in Figure 9.4 and in the map (Figure 9.5) of the whole of South-East England, with some of the fastest rates of growth in towns such as Norwich, Peterborough, Northampton, Telford, Swindon, and Bristol. (Some of these were previously considered as belonging to other regions; London is now gradually becoming their primary centre.) It should be noted that this development is strongly concentrated around certain centres, both new and old; some of the more distant rural areas are still showing strong declines, such as Salisbury Plain.

There is a new phenomenon developing as well: the twice-a-week commuter, who maintains a home with family in a remote part of

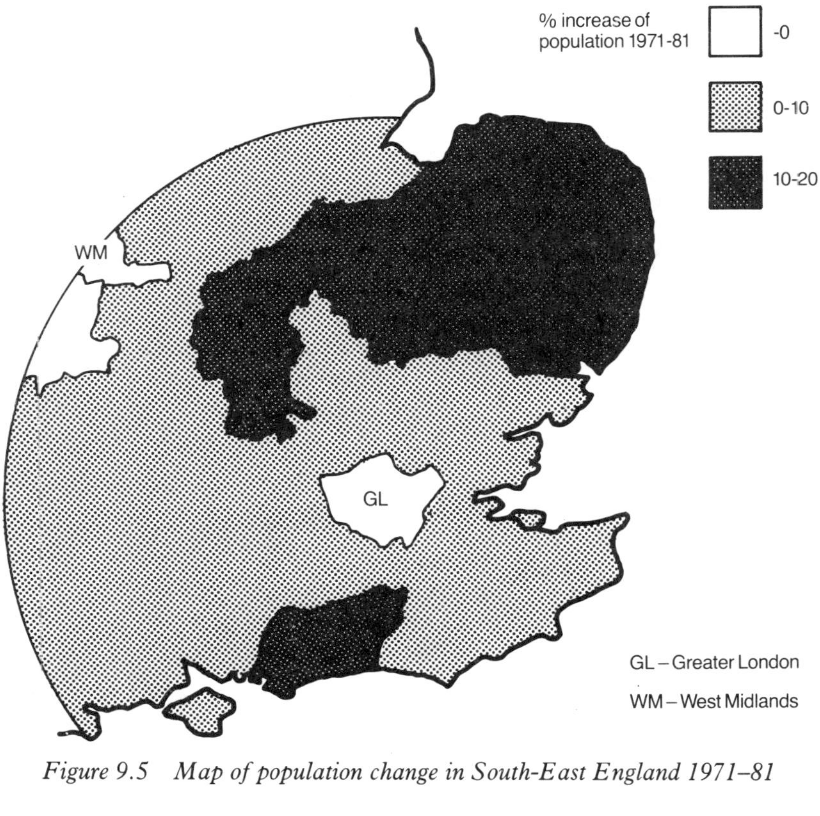

Figure 9.5 Map of population change in South-East England 1971–81

the 'London' area and commutes each end of the week to and from London where he (almost invariably he) lives in a small flat. This is seen most noticeably on the motorways on Monday morning and Friday evening, on the high-speed trains to Exeter, Manchester and Leeds, and even in the air shuttles to places such as Edinburgh, Belfast, Paris and Brussels. But it does not even have to be a five-day week in London now that word processors and data transmission links are common and cheap. No doubt this is the way that the 'London' area will expand as forecast above.

This kind of extension of the 'London' area would no doubt have depressed Clark, who argued in his 1958 paper (*Transport; the maker and breaker of cities*) that a polity should have no more than about 150,000 inhabitants; beyond that size it did not function well as a socio-economic and political unit. Transport, whether by the railway train or the truck, was for him ultimately the breaker of cities, by scattering them over the countryside in ever widening circles of influence. No wonder he spent the last years of his life (up to 1989) in Brisbane, Queensland.

Holroyd (1984) argued that the spread of cities to lower densities, and the spread of population from the cities to the towns and villages of their hinterlands, were both reinforcing, and being reinforced by, the growth of car ownership and use. Because people are moving out to the suburbs and 'ex-urbs', it follows that it is more important to make provision for the new transport facilities they need (and thus for the car) than to improve the transport facilities in the older core of the urban area and thus for public transport. This implicit assumption lies behind much current government thinking. The assumption is mistaken.

It is, in fact, a very old argument, and pre-dates the car. In *Common Sense* (1811), in discussing the changes between the censuses of 1801 and 1811, it is argued that there was a flight of the better-off to the edges of the City, and a replacement of people by warehouses and offices in the centre of the City along the Thames river-front, surrounded by a ring of the highest density tenements for the 'vicious' (sic). The resulting hole of population at the centre of the City which Clark noted was thus already visible by 1811.

This comment went on to note that it was necessary to improve the transport facilities in the ring round the centre to avoid the decay of the tenement area.

> 'It is, however, worthy of consideration, whether the interior of the town does not draw more attention, and there can be no doubt but good streets near the centre of business would be preferred . . . to similar streets in remote parts of the town . . . It is impolitic and senseless to carry the town to Highgate, Hampstead and Clapham, when so bad a use is made of its internal parts; where whole districts consist almost of waste ground, or are occupied by beggary and wickedness'.
>
> *Common Sense* (1811; in 1974 p.205)

Up to the coming of the railways, London was essentially a city maintained by water power. The river Thames was its lifeblood, not only in its very reason for existence at the lowest crossing of the river possible on foot in Roman times, but also as a seaport. The canals from the Midlands brought the food and other commodities to sustain its inhabitants, together with distributor canals around the built-up area. These are shown in Figure 9.6, combining maps from 1746, 1780 and 1823 (Freeman, 1959, Fig. 5).

Highgate, Hampstead and Clapham were at that time no more than small villages, about five miles (eight km) from the centre and thus beyond the coverage of the map.

In 1808, Londoners were treated to the sight of, and invited to ride on, the very first passenger-carrying steam railway train in the world. Richard Trevithick had set up, on a site later to become University College, a circular track round which ran a train pulled by his locomotive 'Catch Me Who Can'.

Whilst this exhibition was not a success (the train was too heavy for the track and only ran for a short while), it was the first public demonstration of the coming revolution that was to sweep the world. Richard Trevithick is rightly honoured by a plaque on the wall of the Civil Engineering Department in Gower Street.

The first railway in London was opened in 1836 between Spa Road, Bermondsey, and Deptford, extending later that same year to its permanent terminus at London Bridge and further out to Greenwich. Spa Road was within a few yards of where my grandfather was to be born in 1866.

But already ideas for taking the new railways right through the centre of London were being floated. Robert Stephenson, who brought the London and Birmingham into its terminus at Euston in

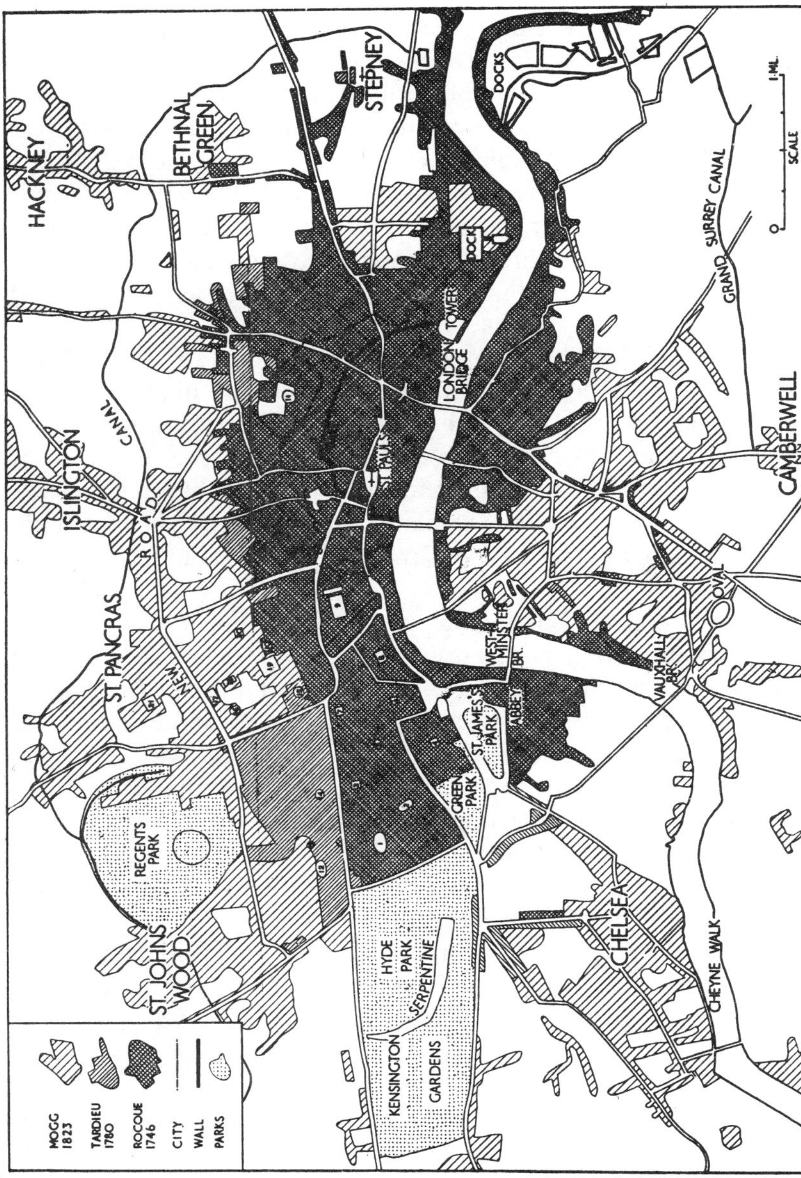

Figure 9.6 London in 1746, 1780 and 1823

1837, said in 1836 that he wanted to continue it southwards under Gower Street, and thus alongside University College, towards the Thames. Charles Pearson, then (in 1837) Solicitor of the City of London Corporation, wanted an underground railway along the New Road (and thus along the northside of University College) and the Fleet valley, to connect up Paddington, Euston and Kings Cross to the City. In fact by the peak of the railway mania in 1844, no fewer than 19 termini were proposed within what became the 'magic oval', or the line joining the major railway termini around the centre of London. It seemed inevitable to many commentators that cross-London links would be built, both from the North to the South and from the East to the West across the centre.

When the Royal Commission set up to investigate all these proposals reported in 1846, it rejected further termini inside this magic oval, and this was enshrined in law. The investigation was thorough, and included origin-destination surveys of traffic to or from the existing termini, their mode of transport and their speeds and costs, so that a cost-benefit analysis of the proposed new termini could be attempted, however rudimentarily, as I have already noted.

It seems that the Commission was heavily influenced by the fact that short-distance traffic from London Bridge to the City tended to use cabs, whereas the longer-distance traffic from Paddington tended to use the cheaper omnibuses, which caused much less congestion, being much more efficient users of road space per passenger. Moreover, freight traffic was far more important then, and the loading and unloading at the termini in the City would have caused considerable congestion.

Despite this prohibition, the railway proposed by Pearson was authorized in 1855, becoming the first underground line in the world, the Metropolitan, when it opened in 1863, initially between Baker Street and Farringdon. It was, of course, steam operated, although the exhaust was blown into a water tank rather than into the passengers' faces, and it was built by cut-and-cover methods. One of its main functions was to bring the traffic from the main lines into the City at Moorgate; therefore, to accommodate the Great Western, which had a seven-foot gauge, it had both gauges of track.

In 1860, a North-South connection was authorized, between Farringdon and a new station being constructed for the London,

Chatham and Dover Railway at Holborn Viaduct, again to bring traffic, from the south, into the City at Moorgate.

By 1862, however, pressure was mounting for a further North-South cross-London connection, and again it seemed inevitable to many commentators. The Cobden Treaty (1860) liberalizing trade with France was expected to increase cross-Channel trade substantially, and the Northern cities required better access to the Channel ports. Colonel Yolland, Inspector of the Railway Department of the Board of Trade, was asked to draw up a plan to connect the London and North West (Euston), Midland (St Pancras) and Great Northern (Kings Cross) lines to the South Eastern (Charing Cross), then being constructed, via a cut and cover railway along new streets which would go past University College and through the notorious slum districts of St Giles and Seven Dials. This railway, the Hampstead, Midland, North Western and Charing Cross Junction, was in fact sanctioned by the Joint Select Committee of the Houses of Parliament in 1864, but was never built because of the financial crash of 1866 and the long depression which followed.

When the extension of the Holborn Viaduct line to Farringdon was opened in 1874, it thus completed the only North-South cross-London line that has been built. A spur at the southern end of Blackfriars Bridge connected it to the South Eastern Railway in 1878. It was, however, used almost entirely by the two railways to operate services from the south into the terminal at Moorgate; no fewer than 19 services were operated in the peak hour (8–9 am) in 1904. The proportion of true cross-London services was small, although it included services such as the Wood Green (Great Northern) to Woolwich, which was withdrawn in 1907.

This line was closed in 1916, but has been reopened on 24 May 1988 as part of the new Thameslink service from Bedford to several destinations in south London, and as far as Brighton. The service is operated by new electric trains. Crucially, however, the line crosses the Thames on a bridge at Blackfriars to link with the elevated south London railways. I mentioned its subsequent upgrading at the end of the last chapter. We can thus argue that it was the coming of the steam railways that enabled the relationship given in Figure 9.2 to have a break-point in 1850. The number of steam train services per day in 1888 is shown in Figure 9.7.

There were two key technical advances that had to be made

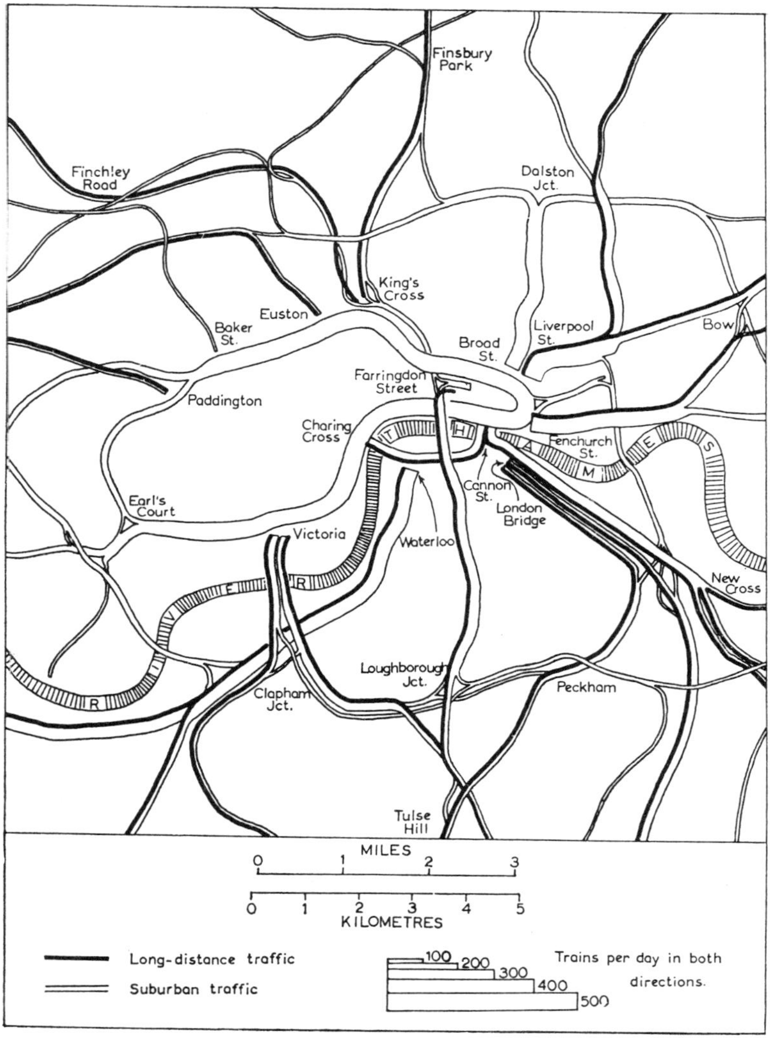

Figure 9.7 The number of trains per day in 1888 in central London (Kemmann, 1892)

before central London could be crossed by rail at a reasonable cost. The first was the development of electric railways. Initially it seemed as if atmospheric railways might be suitable; that is, railways in which the vehicle just fitted inside a tube and was blown or sucked along by air pressure. The Metropolitan had initially considered this, before deciding on steam.

The first Post Office railway, a four foot diameter tunnel between the NW1 District Office (DO) at Mornington Crescent, Euston, Mount Pleasant and St Martin's-Le-Grand, was an atmospheric railway, being opened in February 1863 one month after the Metropolitan. It was, however, plagued with problems of leakages, unlike the system of 2.5″ tubes which the Post Office soon built to link 70 of its Post Offices in central London. Eventually it was converted to electric traction by George Threlfall, but it was too early; the motors did not work well, and it was abandoned in 1880.

Nevertheless, when the Post Office considered a more ambitious scheme for their own railway in 1909, the options were still between pneumatic (atmospheric) and electric railways. They chose the latter; constuction of the nine foot diameter tubes began in 1913 and the line was opened in 1927, between Paddington, the WDO (Oxford Street), the WCDO (Holborn), Mount Pleasant, the ECDO (St Martin's-Le-Grand), Liverpool Street and the EDO (Whitechapel), a distance of 10 km. More ambitious plans to link the rest of the central London termini and the remaining DOs were dropped because of lack of money. This railway was shown in Figure 2.13.

We may yet see vacuum railways, with reduced air pressure, if the full potential of linear electric motors (lem) with magnetic levitation (maglev) are to be realized. Trains powered by rotating electric motors are probably at the limit of their development in, for example, the projected line between Sydney and Melbourne, which is designed for 350 km/h (or 220 miles/h). Now that prototype lem (or maglev) trains can travel at over 300 miles/h (500 km/h) in Japan, air resistance is a limiting factor; but such railways are for inter-city travel, not for travel across the city, although a proposal was made to link the airport at Heathrow to the centre of London by a Japanese maglev train.

The second major advance that had to be made was in tunnelling methods. Attempts had been made to tunnel under the Thames by Ralph Dodd in 1798 and, almost successfully, by Robert Vale and

Richard Trevithick (the same steam engine pioneer) between 1802 and 1808. When the Brunels built the first Thames tunnel—between Rotherhithe and Wapping—it took 18 years from 1825 to 1843. This tunnel was an economic failure until the East London Railway took it over in 1869. Another attempt to build a tunnel, between Whitehall and Waterloo (also for a pneumatic tube), was abandoned in 1866, again because of the financial crash.

It was not until Peter Barlow devised an effective shield, described in his pamphlet *On the relief of London street traffic* of 1867, and James Greathead put it into practice, that tunnels could be dug efficiently; it took just 10 months to dig the tunnel between Tower Hill and Vine Street, Southwark in 1869–70. A 12 passenger cable car service ran for a short time before being removed; it was too near London Bridge to be profitable.

With the problem of tunnelling solved, and with the development of suitable electric motors (the first electric railway in Britain, Magnus Volk's railway at Brighton, opened in 1883 and is still running), the construction of the London subway system could begin in 1888–90. Even then, cable-hauling was considered for the first tube from King William Street to the Elephant and Castle, before electric traction was eventually chosen.

With John Price's development in 1897 of an effective tunnelling machine to replace hand-digging, the Central Line could be excavated faster. Within a few years from 1890, the network of tubes under central London had been completed. Only the Victoria in 1969 and the Jubilee in 1979 have been added since.

It should also be noted that an extensive tramway network had been begun in 1870, first with horse-drawn trams, and then with the development of suitable electric motors the first electric trams began to appear in 1888. These were almost entirely at street level, although the Kingsway development gave the opportunity for one stretch of tunnel.

In welcoming the new tubes in 1890, an editorial in the *Engineer* said (28 November) (the editors were W.H. Maw and J. Dredge):

> 'New lines of streets were formed and fresh channels of communication were opened, but all to little purpose; the crowd of omnibuses, cabs and vehicles of all descriptions in our main thoroughfares remained as dense and impassable as ever.'

Earlier, on 7 November, the editorial had stated:

> 'We look forward to the time when some of our leading thoroughfares will have four deep tunnels laid in them, two for express services and two stopping trains.'

The first express line along the Northern Line was begun in 1939, and terminated later that year on outbreak of war, and never restarted. All that remains are a few deep station tunnels.

In the evidence presented to the Royal Commission on London Traffic of 1903, Sir Henry Oakley, Chairman of the Central London Railway, described the proposals that had been made for cross-London railways between 1855 and 1885 and 1885 and 1903 in plates lxxvii and lxxviia respectively in volume 6. This summary reveals a massive set of proposals.

This Royal Commission on London Traffic was set up under its chairman, Sir Davis Miller Barbour, to inquire into the means of locomotion and transport in London; its terms of reference were to report on:

a) Measures most effectual for improvement by
    i) Development and interconnection of railways and tramways
    ii) Increasing facilities for other forms of transport
    iii) Better organisation and regulation of vehicular and pedestrian traffic.
b) The desirability of establishing some authority or tribunal to which all schemes of railway or tramway construction could be referred.

The Commission deplored the absence of any interconnection between the London County Council (LCC) tramways and those of private companies, and recommended substantial improvements and extensions to the tramway network; that is at street level.

It is, however, for the lucid presentation of the case for cross-London railways by the Advisory Board of Engineers—Sir John Wolfe-Barry, Sir Benjamin Baker and William Barclay Parsons—that this Royal Commission should be remembered. In Volume 7 p.102, the Board states bluntly:

'We are of the opinion that the object to be aimed at in suburban traffic is through running to termini outside "Greater London", and for urban traffic to termini outside the "Central Area".'

Again, on p.115:

'If London could be entirely rebuilt, not only would the streets be made wider and straighter, but the whole system of railway termini would be remodelled. Indeed in most instances probably the termini would be only terminal for a certain portion of the long distance traffic, and the other trains would run through and find their termini in the suburbs.'

And again, in discussing express lines:

'With one station on the outskirt of the City and a very limited number of intermediate ones between those and the terminal stations, it would be possible to provide the artisan dwellers in the suburbs, the passengers on suburban tramways and the population generally with a fast and cheap service of trains.'

The key point for the Board was that termini were inefficient in terms of use of the space and thus increased the operating costs of the railway. (This argument is equally valid today, and was recognised in the decision to reopen the Thameslink service.) Moreover, the depositing of large numbers of commuters at the terminal increased congestion on the streets around the terminal. And the lack of through services limited the choice of residential location for a given workplace. (This latter point was backed up by surveys of travel times and costs to given workplaces.)

The Board was particularly impressed by the schemes of Meik and Beer and of Brereton (who was a partner of Sir John Wolfe Barry) for combined new cross avenues and railways. The former had been presented to the Royal Society of Arts in 1898, the latter was conceived for the Commission. They are given in Volume 6, plates lvii and ciii respectively. Essentially—and inspired by Kingsway—they were new North-South and East-West Avenues, approximately 4.5 miles long, intersecting south east of Mount

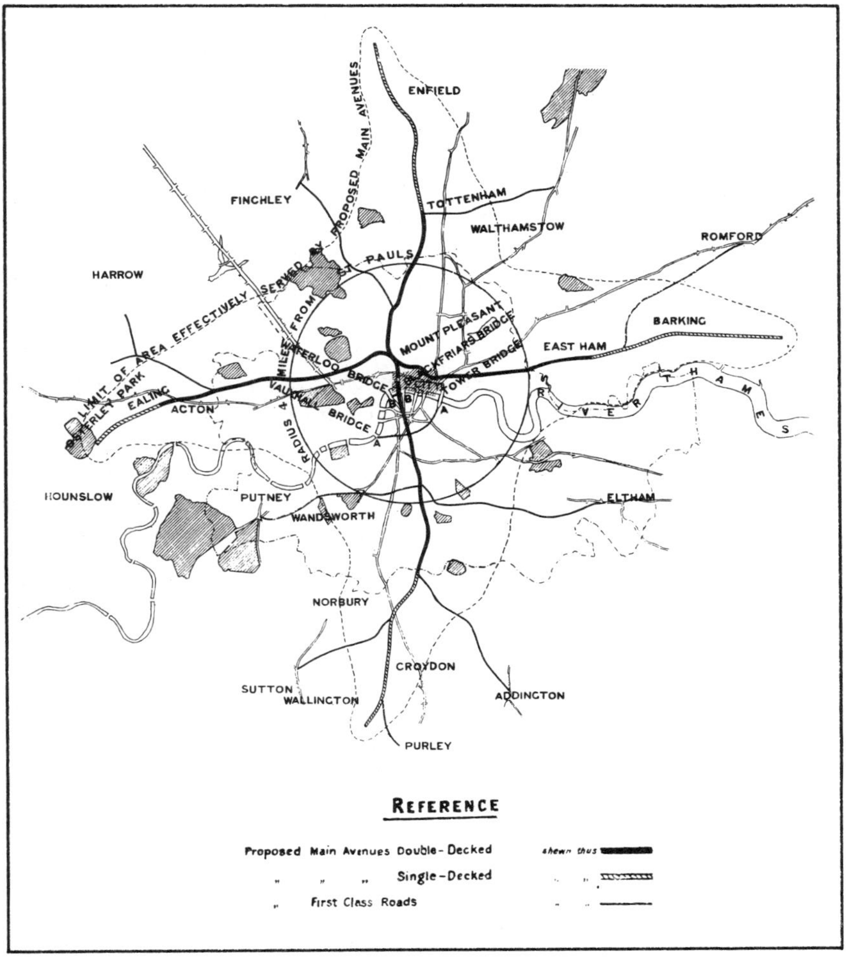

Figure 9.8 Meik and Beer's cross-rail plan (1898)

Pleasant, driven through the poorer areas of the centre (where possible) and whose expense, like Kingsway, would be recouped from the increased value of land. Underneath the Avenues would be railways, both local and express tracks, and at street level would be tramways, again four tracks. Meik and Beer's plan is given above.

The Royal Commission adopted the cross avenue idea as its main structural feature for the new plan; Cross-rail, with local and express services, should therefore be named the Miller Barbour scheme.

The Commission emphasized that:

> 'for the purpose of travelling from the centre of London to the more remote districts, railways are essential, as no other mode of transport gives sufficient speed'.
>
> (paragraph 117, Volume 1, the report)

And again:

> 'no doubt the best and most convenient system for a suburban train service is to have a series of expresses in addition to stopping trains.'
>
> (paragraph 155, Volume 1)

This emphasis on speed is continued in the Commission's next paragraph:

> 'So long as trains are hauled by steam engines, we do not think that much improvement in the speed of suburban train services can be looked for. When electric traction is adopted, the advantage of quicker acceleration will be secured . . . and will enable . . . the journey times of trains to be reduced substantially'.

Unfortunately, the Commission's recommendation for the establishment of a central authority to cover all forms of transport, but on a purely advisory basis, led only to the setting up of a London Traffic Branch within the Board of Trade in 1907; this seemed to feel its main function was to examine road congestion problems rather than the wider field of transport envisaged by the Commission. Cross-rail never got off (or rather under) the ground.

During the period 1901–14, the electric tram and the new motor bus replaced practically all the horse-drawn buses on the streets of central London. Together with the completion, for the time being, of the central London electric tube network by 1908, this accounts for the second break in the relationship given in Figure 9.2 in 1900. The number of trains per day in central London in 1962 is shown in

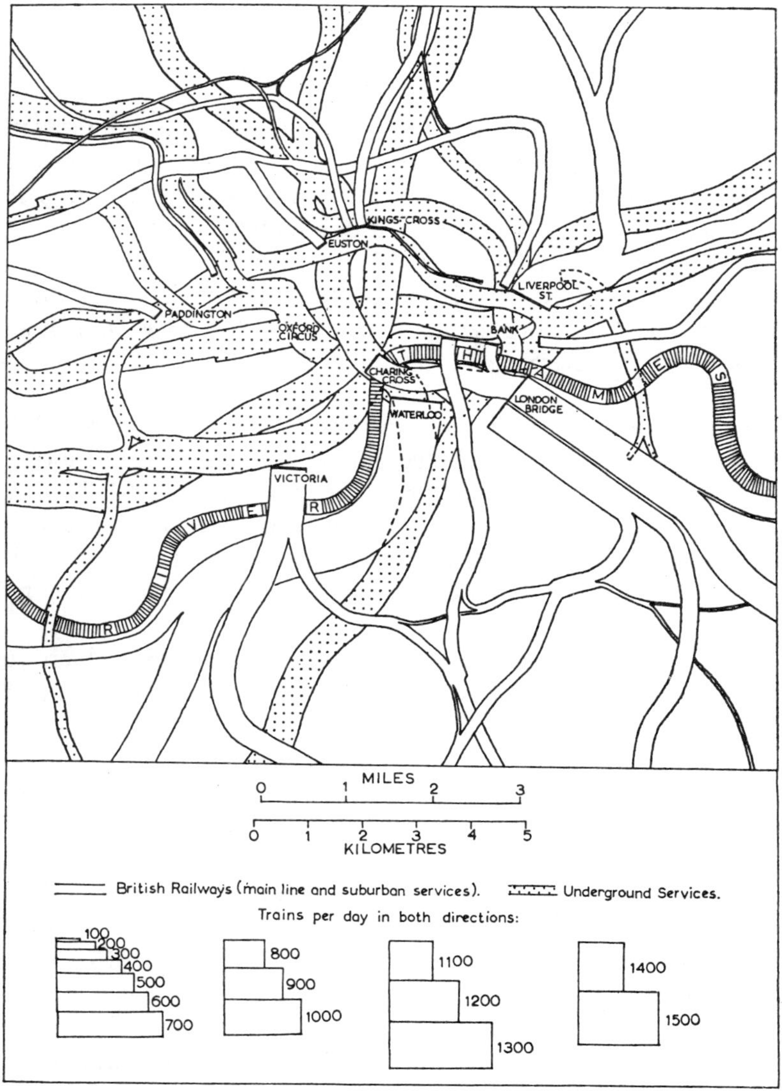

Figure 9.9 Number of trains per day in 1962 in central London (Hall, 1964). Note that this flow scale is 0.4 times that of the earlier map of 1888.

Figure 9.9. The last electric tram ran in 1952; the last steam train in 1962. Central London was at its (temporary) peak of employment (Evans and Crampton, 1989).

One point that stands out very plainly from this map is the way in which the LT tube network carries far more trains on its tracks per day than the BR network. Peak-hour frequencies on the tubes rise to 36 trains per hour (a train every 100 seconds) whereas BR cannot in general manage any higher than half that, owing to the problem of inter-working at terminals. Some terminals with six or eight tracks manage no more services than a double-track tube line, emphasizing the inefficiency noted by the Miller-Barbour Royal Commission.

The next revolution—that of personal motor cars—began rather later in London than in the United States, although (as noted above) motor buses were introduced during the decade from 1904. During the 15 years from 1949, once World War II was over, car ownership quadrupled, and it has since doubled again. Current government forecasts (Department of Transport, 1989a) expect a redoubling by 2025.

This next revolution led to the Buchanan report (1963) which we have already commented on; not a Royal Commission, nor focused on London, but setting out the problem of coping with the revolution just the same.

There is, however, a great difference between this revolution of the internal combustion engine and the earlier ones of steam and electric motors, in that there is no discernible difference, as yet, on the rate of change of the relationship in Figure 9.2. It is worth pondering on why this should be so. To do that we need to look at the characteristics of the way cars are used, and indeed of their ultimate development.

During the next few years, cars will become completely automatic vehicles, true automobiles. Opinions vary as to the rate at which this will happen, and indeed as to the rate at which it will become acceptable; but it is now undeniable that we know enough to make it happen. A number of advances have been made with the advent of computer technology which make this possible.

First, computer control of the engine and power-drive to the wheels—initially to ensure that the maximum economy of fuel was achieved—has now advanced to the stage where the car can drive itself at a speed which is either pre-set or locally imposed by some external control: for example, a speed limit.

Secondly, it is now possible to locate a car in space to within a metre or so, using both navigational satellites out of town or radio beacons inside towns. The communication channels required to cope with the enormous number of messages have expanded in the last few years; partly because of defence demands for location of its personnel and units, partly because of freight companies requiring to optimize their transportation distribution, and partly because of the in-car phone.

Thirdly, it is now possible for a car to find a route through a network of roads with an on-board computer and a map (stored on a single disc) of all the roads in a region, and to display the route, if required, on a screen for the car's occupants to see.

Coupled with this is computer recognition of speech so that all the computer requires is a verbal instruction of the destination, for example 'Take me to work'. Voice-print recognition will prevent unauthorized users from using the car.

Fourthly and most important of all—and no doubt the most contentious—it is now possible for cars to drive safely through a city street, and to park when the destination is reached. Radar or infra-red guidance systems can both be used to assess the safety of the speed and direction of travel. There will always remain one draw-back: that radar shadow will prevent detection of people or objects about to enter the line of travel.

This is exactly the same problem faced by a human driver using sight, with the disadvantage that the human driver takes time to react, does not always react appropriately and may well be driving too fast for the conditions anyway. The automatic driver should thus be much safer than any human driver could be.

It is thus possible that the most compelling reason why such automobiles will be bought and used is not that they free the driver from the moment-to-moment control of the vehicle for possibly some considerable time, but that the horrendous death and injury toll on the roads will become less acceptable. It will be regarded as unsocial behaviour to drive oneself.

Such a shift in attitudes could be encouraged by placing the insurance cost of hospitalization and convalescence of injured persons directly on the car driver rather than on the National Health

Service, together with all the other associated costs of an accident including long-term compensation payments for death and injuries.

It should not be thought that such automation of the car will lead to increased traffic flow in cities. On the contrary, safer driving regimes require lower traffic flows. The Highway Code in the UK describes the safe driving headway for vehicles travelling at a certain speed; these required headways are often ignored, with appalling consequences when accidents happen.

It will, however, eventually lead to increased vehicle ownership above the current estimates of potential saturation levels, since any mobile person will be able to be driven by an automobile, rather than as at present a car driving licence being required before a car is able to be driven.

But as I have argued in previous chapters, an increase in vehicle ownership is not necessarily concomitant with increased vehicle use, but merely with additional choice as to the potential methods of travel available for a particular journey. Whether the car is used or not depends on whether it is the most efficient for that journey.

Table 9.1 Proportion of car journeys by length of journey, Gray 1969

*Length of journey**	*October 1961*	*October 1962*	*October 1963*
0	1	1	1
1	10	10	9
2	14	16	14
3	14	12	13
4	9	10	11
5	8	8	9
6–10	22	21	21
11–15	9	9	8
16–20	4	4	4
21–25	2	2	3
26–50	4	5	5
51+	3	2	2
Total	100	100	100

* *Arrival odometer reading minus departure reading*

Table 9.2 Proportion of car journeys by length of journey Gray (1969)

Length of journey	*Oct 1961*	*Apr 1962*	*Jul 1962*	*Oct 1962*	*Jan 1963*	*Apr 1963*	*Jul 1963*	*Oct 1963*	*Jan 1964*
0–5	56	56	54	57	58	55	51	57	58
6–15	31	29	30	30	30	31	32	29	30
16–25	6	8	8	6	6	7	8	7	7
26+	7	7	8	7	6	7	9	7	5
Total	100	100	100	100	100	100	100	100	100

* *Arrival odometer reading minus departure reading*

If we examine how people actually use their cars we find the following results. The first major study, already quoted in Chapter Seven, was Gray's (1969) survey of *Private Motoring in England and Wales*. In Gray's Table 3.4 we find the distribution of length of car trips for each of the October rounds of the surveys. This is repeated in Table 9.1.

The modal journey (the most frequent journey) length is thus between two and three miles.

If we examine the variation between all the surveys, we find the journey length distributions were as given in Table 9.2.

It can be seen that the values are reasonably stable across the years. There is, however, a bias owing to the omission of holiday mileage in July, which was estimated to add some 10 percent to the average weekly mileage in that month. This would add a correction predominantly to the longer trips.

This series of surveys led directly on to the National Travel Surveys, which include all travel made by household members during the week of the survey. The surveys are, however, not annual but irregular, and have been carried out in 1964 (the pilot) 1965–66, 1972–73, 1975–76, 1978–79 and 1985–86.

We can examine the average speed of car travel by using the 1978–79 National Travel Survey. Table 9.3 gives the distribution of the number of trips, of trip lengths and of trip times in the reported data for car drivers. (Strictly speaking, this is of trip stages—so will include car stages which are part of a longer journey;

Table 9.3 The distribution of car driver trips (NTS 1978-79)

Length of trip (miles)	*% of trips*	*% of length*	*% of time*	*Miles per trip*	*Time per trip*	*Miles per hour*
0 – 1	5	—	1	0.3	3.9	5.0
1 – 2	18	3	7	1.2	7.6	9.5
2 – 3	16	5	8	2.2	9.8	13.5
3 – 5	21	11	15	3.6	14.0	15.4
5 – 10	23	21	24	6.3	20.4	18.5
10 – 15	8	13	12	11.3	29.4	23.1
15 – 25	5	14	11	19.4	43.1	27.0
25+	4	33	21	57.2	102.7	33.4
	100	100	100			
	133600 trips	926400 miles	2614500 minutes	6.93 miles	19.6 minutes	21.3 mph

Table 3.3 1978–9 NTS

for example, by rail.) This can be analysed to give the average speed of travel at each trip length, and the overall average speed.

It may be surprising, but in the UK average trip speeds are only 21.3 miles/h, rising from below 10 miles/h at low trip lengths to 33.4 miles/h for the longest trip lengths over 25 miles (averaging 57.2 miles), with an average trip time of 19.6 minutes and covering 6.93 miles. This includes access time, so is not kerb-to-kerb running time or speed but is measured along the route travelled.

There is a problem here in that the NTS under-records trip mileage by some 10 percent; that is, the sum total of the individual trips over a week is some 10 percent less than the difference between the odometer readings at the beginning and end of the week, but this should not affect the trip distribution. What happens is that in the NTS the survey response falls as the week progresses, until the seventh day when walk trips are included; which means all travel is more accurately recorded again.

If we take an 8,700 miles per year value for the annual distance travelled by the average car, a 6.93 miles per trip gives a number of car trips per day of 3.44 and a travel time per day per car of 67.4 minutes.

In 1983, when I was becoming involved in the journey speed analyses, we already knew that this value was very similar to the value of 68 minutes which we had found (Mogridge and Holden, 1983) for the daily travel time of people in London in 1971 who were only car drivers, and did not use any other mode (it not being possible to identify the separate cars in a household in the GLTS).

It is also worth noting that the average speed of trips by London residents in the NTS is a little slower than the value above, at 18 miles/h (some of which is for journeys outside London). Nevertheless, this value agrees well with the average speed of traffic on London roads, which is also 18 miles/h (Table 2.4).

If we now examine the published tables in the 1985–86 NTS, we find that the speed distribution by length of journey is fairly stable over time. These journeys are where the car is the main mode and include passenger journeys, so are not strictly comparable with Table 9.3.

The speed distribution is not available for the first NTS in 1965.

It can be seen that the speeds are a little different to those given earlier and are somewhat slower on average. It is not clear whether the increase in speeds at the longer distances in the latest survey is a real effect or is a sample effect; one would have to know the average journey length in each of the distance ranges to be sure.

Table 9.4 Journey speed by journey length for car journeys 1972–73 and 1985–86

Journey length (miles)	*mean journey speed (mph)*	
	1972–3	*1985–6*
0 – 1	7.4	6.8
1 – 3	13.2	13.0
3 – 5	17.3	17.3
5 – 10	20.9	22.0
10 – 15	24.9	26.9
15 – 25	26.7	31.3
25 +	33.4	39.3
all journeys	17.8	18.7

para 3.20 1985–6 NTS

Table 9.5 Distribution of lengths of car journeys from 1965 to 1985–86 NTS

Journey length (miles)	*1965*	*% of journeys 1972–3*	*1985–6*
under 5	55	62	61
5 – 10	22	21	20
over 10	22	18	19

The distribution of lengths of car journeys has changed over time, towards shorter journeys. The results in Table 9.3 for journey length distributions are very similar to those in Gray's survey 15 years earlier (given in Table 9.1) except that the proportion of long trips has come down slightly. Table 9.5 may also be compared with Table 9.2.

This is no doubt because, with the increase in car ownership, cars are now available for many more of the non-work trips which tend to be shorter.

Cars are thus filling a niche of travel demand which is typically fairly short-distance, fairly low-speed and, in the UK at least, mostly within urban areas. Cars are essentially replacing walk and bus travel.

They are as a consequence having no effect *per se*, on the rate at which the city expands. That can only be solely due to the railway system.

Over the next few decades, I expect the gradual introduction of high-speed trains capable of 140 miles/h (220 km/h), the current BR target, to continue. This process will be enhanced by the growing realization in Europe that the skies are becoming too congested and that the trunk-haul routes between the major European cities must be by high-speed rail, leaving the airlines to cover the smaller and longer inter-city routes.

At the intra-city or commuting level, these high-speed trains should be connected across the centres of the cities. Cross-rail connections between all the commuter lines should be built over the next 50 years and the commuter terminals replaced by underground stations, just as the Miller-Barbour Royal Commission suggested and for the same reasons (which are just as valid today and for the future as nearly a century ago). A start has been made with the Thameslink service, and its success should lead to a realization that this is the way forward for the BR rail system.

Whilst this will substantially increase the accessibility of the conurbation, and particularly of the centre, we are as yet unable to say how it will affect the way in which the relationship in Figure 9.2 will change as a result; that is, the rate at which the city will spread out into its region.

10 *Road Building*

Some fallacies

Those who argue that roads must be improved in London seem to fall very easily into a number of traps or fallacies.

i) London is grinding to a halt

This is a very old argument. Essentially, people who use this argument take a limited time series of the traffic speed data. They start in 1974–76 when the traffic speed was unusually high because of the petrol price rise, and difficulty of obtaining petrol. They finish in 1983–86 when the traffic speed was unusually low, possibly because the increase in demand for travel to the central area had so increased the load on the public transport system that the running speed of the trains had reduced owing to congestion at stations; thus, through the equilibrium process, the traffic speed on the roads was reduced.

A regression equation through this limited data on traffic speed thus appears to show that speed is slowing down, and that by extrapolation it will be even slower in future. A value of 7 miles/h by 2000 is often quoted.

The regression of the morning peak speeds from 1974–86 gives:

$$\text{traffic speed} = 14.15 - 0.59t$$

(where t = 0 in 1974, and in 3 years units) This gives a traffic speed of seven miles/h in 2010. The argument is that roads must be improved to prevent this slow-down.

This argument is fallacious because it does not take a sufficiently long time span of data into account. One might as well argue that from 1962 to 1974 (the same timespan) the data showed that road speeds were increasing.

ii) London road speeds are constant; as traffic is increasing we must build more roads otherwise road speeds will fall in future

In its modern form this argument has been put most clearly by the *Wheels for Change* pamphlet of the Movement for London, December 1987. The argument used to be used entirely in the central area, but now that it has been realized that the North Circular Road all-day traffic speeds have been constant for 50 years it is being used much more widely. A recent example was the letter to the *Times* on 4 June 1988 by the Chairman of the Confederation of British Industry, John Banham.

The argument is essentially that car ownership has grown substantially—in fact in the 50 years from 1936 to 1986 it has grown nationally from about 1.8 million cars to about 18 million cars—and that it will continue to grow fast. Current growth rates are about 5 percent per annum. Moreover, the amount of freight to be moved by road is also growing fast. National surveys show that the average total distance travelled by each car per year is fairly constant at about 8,700 miles, so logically it would seem to follow that we need more road space to prevent a slow-down in traffic speeds.

This argument ignores the fact that the increase in road space in London has not kept pace with the increase in car ownership in the past. To take the specific example of the North Circular Road, it is certainly not 10 times as big in capacity as it was in 1936. Why, then, have road speeds not fallen already, and substantially, given the sensitivity of speed to flow in congested conditions?

The simple reason is that cars based in London do not travel entirely in the London area. To take perhaps an extreme example, the car I share with my wife was used for about 1,000 miles last year, almost all of which was done outside the London area—or on direct routes in or out of the London area since we live in the centre.

The vast majority of car trips in London are of relatively short length, short duration and slow speed. The increase in traffic by these cars takes place outside the London area.

It can also be argued that there is a degree of peak-spreading (i.e. that more travel is now taking place outside the peak hours) but it must be remembered that the speed surveys on the North Circular Road are all-day averages. Even if more travel is outside the peak

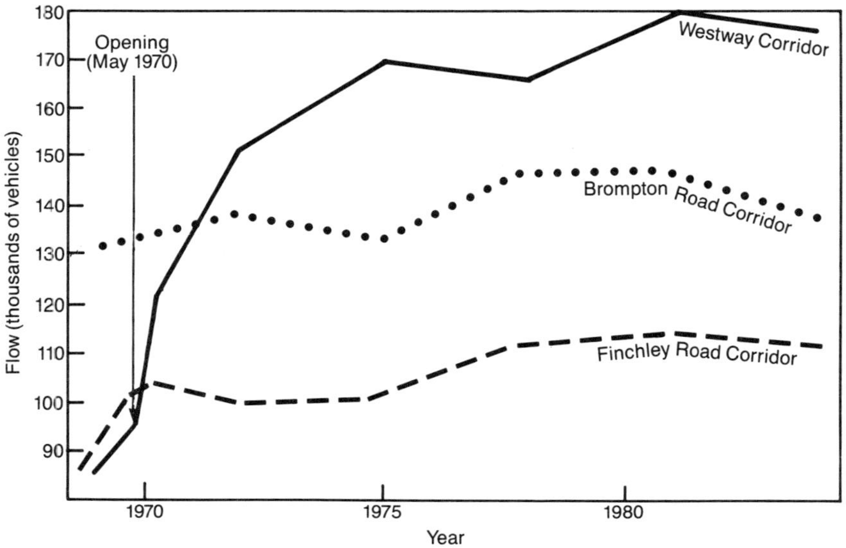

Figure 10.1 Traffic flows on the M40 and adjacent corridors

hours, it is still travelling at a relatively slow speed; the peak speeds on the North Circular are 19 miles/h, the same as the average speed in the whole area in which the road lies; whereas the evening off-peak speed is 28 miles/h, only about 50 percent higher.

Thus the vast increase in car ownership over the last 50 years has not led to a reduction in the all-day average speeds on roads such as the North Circular, so there seems to be no reason to suppose that it will do so in the future as car ownership continues to grow. It seems to be quite absurd to claim that road engineers have just built exactly the amount of road space required to keep speeds constant, since the amount of road space built bears no relation to the increase in car ownership. Moreover, in cases where road space has been reduced for cars, such as the conversion of Oxford Street to a bus and taxi-only road in 1972, there has been no diminution of the average speed of road traffic in the surrounding area; although there were indeed horrendous jams whilst the system settled down to a new equilibrium, over a period of about six weeks.

Even more telling, in cases where a substantial increase of capacity has been provided, it can be demonstrated that an increase in road capacity has no effect whatsoever on the volumes of traffic being handled on parallel roads, after the system has settled down to a new equilibrium. For example, Beardwood and Elliott (1985) analysed, among others, the case of the opening of Westway (the M40) between White City and Edgware Road/Marylebone Road in 1970; improving the capacity in the radial corridor of the A40 from the north-west to the centre of London. In the adjacent corridors—Brompton Road, the A4, to the south and the Finchley Road, A501 and A5, to the north—there was no diminution of traffic whilst the M40 built up to its daily load of 80,000 vehicles per day over a period of 10 years. This is shown in Figure 10.1.

All that increases in road capacity do is allow people to abandon public transport in favour of the car.

iii) Even if road speeds remain unaltered by increases in capacity, reliability of journey times is improved

This argument has resurfaced in a letter to the *Times* from Olaf Lambert, Chairman of the British Road Federation, on April 27. 1988. He believes:

> 'The only realistic way to give Londoners predictable journey time and reliable delivery of goods and services is to improve the roads.'

His assertion that improved roads lead to increased reliability of journey time flies in the face of the evidence. The evidence available suggests that reliability—or the coefficient of variation of journey time—is strongly dependent on journey speed, higher speeds giving lower coefficients of variation and thus higher reliability. The basic data was discussed in the work of Wardrop and Turner (1951), and there is no later evidence which would cause a modification of their conclusions.

If improving the roads cannot change traffic speed, for the reasons discussed above, then likewise it cannot change the reliability of journey times.

iv) Even if traffic speeds remain unaltered by improving the roads, we must do it because people prefer to use their cars rather than travel on public transport

In the same letter quoted above from Olaf Lambert, he finishes with the assertion:

> 'We can create a better transport system for London, but it must accommodate the Londoners' wish to own and use cars in the same way as that right is enjoyed throughout the rest of the country.'

I have shown above that the use of the car in London, while at slightly lower speeds on average than in the rest of the country, is not so dissimilar as many people seem to believe. The fundamental point that Olaf Lambert is making, however, is that once a car is purchased the owner is entitled to be able to use it.

I have no wish to discourage the ownership of cars, and indeed in my own work have forecast much higher levels of car ownership than the government are presently contemplating (Mogridge, 1989). It seems to me, however, to be quite absurd to spend enormous amounts of money on road construction if all that is achieved is the same journey speed as before, and to run the risk of a lower journey speed if public transport services are reduced, when for the same amount of money spent on investing in public transport one can achieve a speeding up of both the public transport system and the private car, for those who still choose to use it.

The individual cannot choose by himself a better public transport system; that can only be done through government decision. But it is absurd to claim that because the individual can choose to buy a particular car, then the government must provide him with the wherewithal to travel at high speed in it.

It is often argued that, because the individual has paid a great deal of tax in purchasing and using the car, the individual has a right to expect the government to spend that tax money on the roads. This argument, that of hypothecation, is peculiar to roads and is not used, for example, to justify expenditure on pubs for alcohol consumers, or smoking rooms for smokers, or sound rooms for hi-fi addicts, all of which are highly taxed.

The responsibility of government is to allocate the resources of society in such a way as to maximize the welfare of the people, and if public transport investment is the efficient way to do that then that is the way that resources must be spent, even if those resources come in part from tax on the private car.

After all, what the individual wants is mobility, travel at higher speed in comfort and security. It is the stated duty of government to provide that. It cannot be done in London by improving the roads. It can only be done by improving the public transport system. Therefore, we have to design the public transport system to provide the maximum increase in journey speed for the amount of resources available.

v) Congestion costs Londoners £1.5 billion per year; therefore more roads must be built to relieve congestion

This argument goes back a very long way, as already noted above. In the latest pamphlet from the British Road Federation, (*The Cost of Congestion*; June 1988), the details of their calculations are given as follows. The costs of congestion are measured from the difference in journey time under 'ideal' conditions and in reality, multiplied by some value of time, plus the additional vehicle costs of travelling slower. The definition of 'ideal' condition is therefore crucial to the argument.

The definition chosen is that condition where the driver is unimpeded by any other driver, in effect where he is travelling in the conditions of the dead of night (say at 3 am), where there is no other traffic to speak of. One can call these 'Ghost town' conditions. In the centre of the city, the speed assumed for these ghost town conditions is 37.65 km/h (23.38 miles/h); in inner London 42.48 km/h (26.38 miles/h) and in outer London 54.70 km/h (33.96 miles/h). The values of the free-flow speeds are taken from Glaister (1982). These speeds are approximately twice the speeds actually achieved (quoted as 19.35, 19.36 and 32.16 km/h respectively for 1985/86), so the congestion cost is therefore approximately half of what people spend on road travel.

These speeds actually achieved are, however, *not* the speeds actually achieved but the BRF's estimate of what the speeds would be in 1985/86, based on the speeds in 1980/81, applying a growth factor of 'number of vehicles licensed' to the intervening years from GLC-wide data and using the speed-flow relation given by Glaister to estimate speed. These estimates of speed may be compared with the actual speeds in 1984/86, as given above, and repeated here for convenience. (I have used a simple weighting of 3:6:3 for the morning peak: midday off-peak: evening peak, rather than flow weighting.)

	BRF estimate	*86/8*	*actual* *83/5*	*80/2*
Central	12.02	11.1	11.8	11.9
Inner	12.03	12.3	14.1	14.9
Outer	19.98		20.2	20.4

It can be seen that the BRF method has substantially underestimated inner London speeds in 1983/85 though the later speeds have fallen slightly. Curiously, however, given the method it has not markedly underestimated the central and outer London speeds, which have not changed appreciably in the period. Obviously one must re-examine Glaister's data.

The actual BRF method, whereby future speeds are estimated using growth factor methods on vehicle licences and speed-flow relationships, will overestimate the costs of congestion because it will underestimate speeds. Speeds are not set in this way by a speed-flow relation when demand is suppressed; they are set by the equilibrium with public transport journey speeds.

There are indeed people for whom the ideal conditions would be if everyone could travel at the speed they could on an unimpeded motorway—i.e. at 70 miles/h for the majority of their journey—and only at the ends of their journey travel at the local speed. As we have seen in Chapter Nine, the vast majority of journeys by car are actually very short, and certainly comparable in length to the distance between intersections on a motorway operating at 70 miles/h. The vast majority of trips could therefore never be on a motorway network, but would always be on a local network.

Implied therefore in these definitions of 'ideal' conditions is that it is possible to build a road network on which everyone can travel as if there were no other driver also travelling at the same time. These are patently absurd conditions because the problem of urban travel is that we *do* all want to travel to the same location at the same time. It is in the very nature of a city that that should be so, at least in our present culture where face-to-face contact is all-important.

The cost of congestion is therefore an invalid concept in an urban area like London. What we need to do is to consider what are the ways available to speed up movement, and how much it costs to do so; not to attempt to define an 'ideal' condition which is impossible to achieve.

We cannot reduce congestion by building more roads since immediately we get more traffic to fill them up to the same speed as before. The only way to reduce congestion—i.e. speed up traffic—is to introduce better public transport facilities which reduce the number of people who travel by car on the roads.

11 Conclusions

In this book I have attempted to do two things: to show that road speeds in the centres of large towns are very stable over long periods of time (I use London, my home base, as an example); and to show how I think they can be improved.

I have therefore examined the various sets of data on travel speeds available in London, which go back as far as 1863, to demonstrate this stability. I conclude that road speeds were lower in the last century, albeit on the sparse evidence available; and substantially improved once London was provided with an electric tube system across its centre during the period 1890 to 1908.

The reason appears to be very simple. We assume that people travel by the best method available to them. They will therefore prefer to travel by individual vehicles until the journey attractiveness becomes as bad as that by communal vehicles. The journey speeds using communal vehicles therefore set the journey speeds using individual vehicles if there is not enough space for everyone to travel by individual vehicles.

Individual vehicles are very inefficient users of space, both moving and stationary, whereas communal vehicles are more efficient the more passengers they have and the less space they use for stops. It has long been recognized that it is impossible to cater for all movements by individual vehicles in the centre of a city like London; this was the key concept behind Buchanan's *Traffic in Towns* in 1963.

Buchanan, however, merely commented that all those who were unable to use their preferred method of travel would have to be provided with good and cheap public transport. It is possible to go much further than this and specify exactly what is meant by good and cheap.

To do this I have examined the development of transportation modelling in London since the first detailed network models became possible with the advances in computer technology in the 1960s.

Initially, the first such models were based on forecasting road traffic demand on a network of new motorways operating at design speeds, with a very rudimentary treatment of public transport. It soon became obvious, however, that no possible road network could provide for the anticipated future demand levels; the roads option was decisively rejected by the electorate in London in 1973, and a policy of subsidizing public transport operating costs adopted instead.

Attention switched to road pricing in the modelling in the 1970s in an effort to estimate the price needed to restrain traffic demand to an optimum level in congested areas such as central and inner London. Whilst two consecutive studies showed some of the benefits which could be achieved by a policy of road pricing, the ignoring of the fundamental economic theory behind the pricing of road traffic led to unsatisfactory conclusions and a failure to adopt the policy.

Whilst the theory goes back to Dupuit in 1844, the major advance comes from Marshall's work in 1890, and its application to roads by Pigou in 1920 and Knight in 1924. Knight argued that there was a price which could be levied on users of a congested road which would reduce the consumption of resources by the community but leave no-one any worse off than before, provided users who did not wish to pay the price could switch to an alternative whose costs were in equilibrium with the congested road.

Edgeworth, a year later, demonstrated that it was in fact possible for all users of alternative facilities to be better off after a price had been levied on users of one of them, though it was left to Hotelling in 1932 to generalize the conditions under which that was so. Essentially, users of a road where costs increase with increasing flow should have a tax levied on them, and users of public transport where costs decrease with increasing flow should be given a subsidy; so that the users are made to perceive the marginal costs of their travel and therefore pay exactly for the consumption of resources involved in their journey. Under these conditions, the maximum net social benefit will be achieved and everyone will be better off.

More specifically, road users in congested conditions are being subsidised by the community because they are paying too low a price for the resources that they are consuming; a tax must be levied to offset this subsidy. Public transport users on the other hand are being taxed by the community because they are paying too high a price for the resources that they are consuming; a subsidy must be levied to offset this tax. If this is not done, too many cars will be using the road space and too few buses and so on, giving much slower speeds than could be achieved if prices were set at resource costs.

Whilst the 1970s saw demonstrations that public transport costs did indeed show economies of scale (a fact which had in fact been discussed for well over a century), it was another aspect of the equilibrium between individual and communal travel that first surfaced in my own work on the study of journey speeds, the Downs-Thomson paradox. Downs in 1962 and Thomson in 1977 realized that if road capacity is increased in congested conditions and draws travellers off the public transport system, the attractiveness of the road system after the capacity increase could be worse than that before, owing to the shift in the equilibrium. Conversely, only by improving the public transport system can the attractiveness of the road system be improved. This seemed to me to be a better explanation of the very slight deterioration in journey speeds during the 1960s and 1970s in London than any other.

During the 1980s, transportation modelling in London has at last incorporated the decreasing costs of bus and rail operation with increasing flows. The most recent studies conclude that road building in the congested conditions of the central and inner city is a policy with a negative social benefit, whereas a policy of creating new public transport infrastructure in those areas has a positive social benefit. More particularly, a policy of road pricing coupled with a policy of subsidy to public transport operating costs gives the largest net social benefit for a given network.

There is still the issue of optimising the system to be tackled, since the marginal costs of journeys by individual transport and the various forms of communal transport are not as yet well defined, especially in the different densities of urban development in London. It is already quite clear from sensitivity testing, however, that these conclusions are robust.

Most transport models up to now have been concerned with optimising the travel demand when the traveller chooses on the basis of perceived costs; that is choosing the journey that is most attractive. But in 1952, Wardrop had specified that there were two criteria that could be used to specify journey choice: that where the user minimised costs, and that where costs were minimised over the system. The former—the user-optimised travel demand—can only be equal to the latter—the system-optimised travel demand—if a directing intelligence imposes the appropriate taxes and subsidies to ensure that users pay the marginal costs of their journey and thus for the resources that they consume.

Van Vliet and Dow in 1979 have already given the conditions for achieving a system-optimised road travel demand; Bly and Oldfield's (1986a) work on public transport costs has now achieved the same for public transport. What is needed now is to apply these two advances to the transport models and in particular to the design of new transport infrastructures to see which have the greatest benefit/cost ratios when travel costs are set at their marginal levels.

I suspect that it will be shown that the commuter rail system of London should be taken right through the centre and the commuter terminals abolished and replaced by underground stations. I suspect that some of these links should be for regional express railways operating at up to 150 miles/h or more as London expands. I suspect that it will be shown that the non-radial routes of the rail system in inner London should have an up-graded metro-type service. Some of this construction will take the next 50 years.

I suspect that it will be shown that some road space on the major roads in central London and on the inner London radial routes should be devoted to the provision of a high quality segregated tram network, and indeed in other strategic centres in London; on other trunk routes, continuous bus lanes would be more appropriate. I suspect that it will be shown that a network of small minibuses would be far more cost-efficient than the large buses for serving routes other than trunk routes and possibly for providing express services at a premium even on trunk routes. Most of this adaptation could be started immediately.

I have concentrated on London because it is one of the largest world cities, because its traffic problems are typical of the complexity of such cities and because there is a long history both of data collection and of attempts to understand the traffic problem by analytical methods.

To the extent that the use of space is a common problem in all cities, then the lessons learned in London will be applicable to all cities, though necessarily the scale and magnitude of the problems may be different.

The costs of the various forms of transport may well be different, and indeed will change at different rates through time, as technology advances.

The key point that we have learnt in London is that only by ensuring that travellers are faced with and pay the true costs of the resources that they consume on each journey can an efficient pattern of transport supply be devised which will enable all the multifarious travel demands to be met at the lowest cost in resources.

It does not have to be:

> 'Jam yesterday, jam today and jam tomorrow.'

We know now how to speed up travel, and thus traffic, in towns.

Bibliography

P. Abercrombie, *Greater London Plan 1944;* London, HMSO, 1945.

W.M. Acworth and W.T. Stephenson, *The Elements of Railway Economics*; chapters 3–5; Oxford, Clarendon Press, revised edition, 1924.

R. Ajo, 'London's Field Response in Terms of Population Change'; *Acta Geographica* I: 18(2), II: 18(3) (1964).

R. Ajo, 'On the Structure of Population Density in London's Field'; *Acta Geographica*: 18(4), 17pp (1965).

T. Anderton, *Comparison of the District and Zonal Level Assignments in the GLTS Model*; London, Greater London Council Department of Planning and Transportation, 1978 (Unpublished).

J. Banham, Letter to *The Times,* 4 June 1988.

T.C. Barker and M. Robbins, *A History of London Transport: Passenger Travel and the Development of the Metropolis*; 2 vols., London, George Allen and Unwin for the London Transport Board.

R.E. Barkley, 'An Annotated Bibliography on Traffic Volume Studies and Origin-destination Surveys' (unpublished M.Sc. thesis, Purdue University, 1950).

Sir D. Barran, *The London Rail Study*; London, Greater London Council/ Department of the Environment, 1974.

D. Bayliss, *Area Licensing in Central London. Proceedings of Conference on Urban Transport and the Environment 11*; 253–78, Paris, Organisation for European Cooperation and Development/European Conference of Ministers of Transport, 1979.

J.E. Beardwood, Private communication (work to be reported as part of an Economic and Social Research Council project), South East Regional Research Laboratory, London School of Economics, London, 1989.

J.E. Beardwood and J. Elliott, 'Roads Generate Traffic'; *Highway Appraisal and Design,* (1989): 270, 97–109, London, PTRC Education and Research Services Ltd for Planning and Transportation Research and Computation (International) Co. Ltd.

M. Beckmann, C.B. McGuire and C.B. Winsten, *Studies in the Economics of Transportation*; New Haven, Connecticut, Yale University Press for Cowles Commission for Research in Economics, 1956.

M.E. Beesley, 'The Value of Time Spent in Travelling: Some New Evidence'; *Economica*: 32 (1965), 174–85.

M.E. Beesley and G.J. Roth, 'Restraint of Traffic in Urban Areas'; *Town Planning Review*, 33, (1962), 184–96.

M. Ben-Akiva, N. Litinas and K. Tsunokawa, 'Continuous Spatial Choice: the Continuous Logit Model and the Distribution of Trips and Urban Densities'; *Transportation Research*: 19A (1985), 119–54.

J.H. Blase, 'The 1980 Recalibration of the GLTS Model'; *Transport Studies Note 97*: London, Department of Planning and Transportation/Greater London Council, 1980. (Unpublished.)

J.H. Blase, 'Radial Journey Speeds to Central London 1962–1981'; *Transportation Studies Note 171 (GLTS Analysis Report 14)*: London, Department of Planning and Development/Greater London Council, 1985. (Unpublished).

H. Bleicher, *Statistische Beschreibung der Stadt Frankfurt am Main and Ihrer Beölkerung. J.D. Sauerläder's Verlag*, Frankfurt am Main, 1892.

D.E. Blumenfeld, 'Effects of Road System Design on Congestion and Journey Times in Cities' (unpublished Ph.D. thesis, University College London, 1972).

H. Blumenfeld, 'On the Concentric Circle Theory of Urban Growth'; *Land Economics*: 25 (1949), 209–13.

P.H. Bly, R.H. Johnston and F.V. Webster, 'A panacea for road congestion?'; *Traffic Engineering and Control*: 28 (1987), 8–12, 19-20. See also: M.J.H. Mogridge and D.J. Holden (1987) and R.H. Johnston (1988).

P.H. Bly and R.H. Oldfield (1986a), 'An Analytic Assessment of Subsidies to Bus Services'; *Transportation Science*: 20, 200–12.

P.H. Bly and R.H. Oldfield (1986b), 'Structure and Calibration of the London Area Model' (unpublished working paper, Special Research Branch, Transport and Road Research Laboratory, Crowthorne, Berks).

D.J. Bogue, *The Structure of the Metropolitan Community: A Study of Dominance and Subdominance*; Ann Arbour, Michigan, Horace J. Rackham School of Graduate Studies, University of Michigan, 1949.

Sir C. Bressey, 'The City'; a film talk. Production: Cavalcanti (Central Office of Information?), London, 1939.

Sir C. Bressey and Sir E. Lutyens, *Highway Development Survey 1937 (Greater London)*; London, HMSO, 1938.

British Road Federation, *The Cost of Congestion*; London, BRF, 1988.

C. Buchanan, *Traffic in Towns: A Study of the Long-term Problems of Traffic in Urban Areas*; London, HMSO, 1963.

Buchanan, Colin, and Partners, see: The MVA Consultancy (1988) (1989).

C.M. Buchanan, *London Road Plans 1900–1970. Research Report 11*; London, Intelligence Unit/Greater London Council, 1970.

J.M. Buchanan, 'The Pricing of Highway Services'; *National Tax Journal*, 1952: see also Buchanan (1956).

J.M. Buchanan, 'Private Ownership and Common Usage; the Road Case Re-examined'; *Southern Economic Journal*: 22 (1956), 305–16.

P.A. Buckles, 'Journey Times by Bus and Car in Central London in 1972'; *Traffic Engineering and Control*: 15 (1973), 337–39.

R. Bussière, *Morphologie Urbaine, Repartition de la Population*; Paris, Centre de recherche d'urbanisme, 1968; (in English: The spatial distribution of urban population), The Hague, Fédération Internationale pour l'Habitation, l'Urbanisme et l'Aménagement des Territoires, 1968.

R. Bussière and T. Stovall, *Systèmes Évolutifs Urbains et Régionaux à l'État d'Équilibre*; Paris, Centre de recherche et de rencontres d'urbanisme, 1981.

L. Carroll, *Through the Looking Glass and what Alice Found There*; People's edition, London, Macmillan, 1898.

E. Casetti, 'Alternate Urban Population Density Models; an Analytical Comparison of their Validity' in *Studies in Regional Science*; Ed. A.J. Scott. *London Papers in Regional Science,* 1, 105–16, London, Pion, 1969.

I. Catling and B.J. Harbord, 'Electronic road pricing in Hong Kong: 2) the technology'; *Traffic Engineering and Control,*: 26 (1985) 510–15.

Central London Rail Study. A joint study by the Department of Transport, British Rail Network SouthEast, London Regional Transport and London Underground Ltd., London, Department of Transport, 1989.

C. Clark, 'Urban Population Densities'; *Journal of the Royal Statistical Society*: 114A (1951), 490–496.

C. Clark, 'Transport: maker and breaker of cities'; *Town Planning Review*: 28 (1958), 237–250.

C. Clark, *Population Growth and Land Use*; London, Macmillan, 1967.

'Common Sense', original communication in *The Monthly Magazine*: 31 (1), 1811; reproduced as 'Economic Base and City Size; an 1811 Commentary on London' in *Land Economics*: 50 (1974), 202–205.

R.D. Coombe, *The Inner London Bus Priority Model. Research Memorandum 416*; London, Department of Planning and Transportation/Greater London Council, 1973.

J. Craig, 'Population Density and Concentration in Great Britain. 1931, 1951 and 1961'; *Office of Population Censuses and Surveys, Studies on Medical and Population Subjects, No 30*: London, HMSO, 1975.

J. Craig, 'Population Density and Concentration in Great Britain. 1951, 1961 and 1971'; *Office of Population Censuses and Surveys, Studies on Medical and Population Subjects, No 42*: London, HMSO, 1980.

M. Dasgupta, M. Frost and N. Spence, *Mode Choice in Travel to Work in British Cities 1971–1981*; (research paper, Department of Geography. London School of Economics. London, 1989).

S. Daspal and T.R. Gopalakrishnan, *Journey Time in Central London. Transport Network Theory Unit. Paper 98*; London, London Business School, 1969.

J. Dawson, 'Electronic road pricing in Hong Kong: the pilot stage'; *Traffic Engineering and Control*: 24 (1983), 372–74. See also: Dawson and Brown (1985), Catling and Harbord (1985), Harrison (1986), Dawson (1986), Harrison, Pell, Jones and Ashton (1986).

J. Dawson, 'Electronic road pricing in Hong Kong: 4) conclusions'; *Traffic Engineering and Control*: 27 (1986), 79–83.

J. Dawson and F. Brown, 'Electronic road pricing in Hong Kong: 1) a fair way to go?'; *Traffic Engineering and Control*, 26 (1985), 522–29.

R.F.F. Dawson and J.G. Wardrop, *Passenger-Mileage by Road in Greater London. Technical Paper 59*, Road Research Laboratory; London, HMSO, 1962.

Department of Transport, *Traffic Appraisal Manual*; London, Department of Transport, 1981.

Department of Transport, 'London Traffic Monitoring Report for 1987'; *Statistics Bulletin (88) 39*; London, Department of Transport, 1988a.

Department of Transport, *Observations on the House of Commons Transport Committee Third Report, Session 1986–87,*; Cm. 420, London, HMSO, 1988b.

Department of Transport, *National Road Traffic Forecasts (Great Britain) 1989*; London, HMSO, 1989a.

Department of Transport, *Traffic Speed on Roads in Central London: A Compilation of Historical Results. Transport Statistics Report*; London, Department of Transport, 1989b.

Department of Transport, 'Traffic Speeds on London Roads: Inner area survey'; *Statistics Bulletin (89) 30*; London, Department of Transport, 1989c.

T.A. Domencich and D. MacFadden, *Urban Travel Demand*; Amsterdam, North Holland, 1975.

A. Downs, 'The Law of Peak-hour Expressway Congestion'; *Traffic Quarterly*: 16 (1962), 393–409. Reprinted in A. Downs, *Urban Problems and Prospects*; Markham, Chicago, 1968.

A. Downs, 'The automotive population explosion'; *Traffic Quarterly*: 33 (1979), 347–62.

N.C. Duncan, A.W. Christie and M. Marlow, 'Traffic speeds in towns: further analysis of the urban congestion surveys'; *Traffic Engineering and Control*: 21 (1980), 576–79.

J. Dupuit, 'Annales des Ponts et Chaussées'; 2nd series, vol. 8 in di M. di Bernardi and L. Einaudi, *De l'utilité et de sa mesure*; Turin, La Riforma Soziale, 1932. In English: On the measurement of the utility of public works, trans. R.H. Barback in *International Economic Papers*: 2 (1952), 83–110 and in Goodwin and Jones (1989).

East London Rail Study A joint study by the Department of Transport, British Rail Network SouthEast, London Regional Transport and London Underground Ltd; London, Department of Transport, 1989.

F.Y. Edgeworth, *Papers Relating to Political Economy*: 1 (1925) 132, 143ff and 2, 401, London.

B. Edmonston, *Population Distribution in American Cities*; Lexington, Maryland, D.C. Heath, 1975.

Engineer, The; Editorials 7 November and 28 November 1890, eds. W.H. Maw and J. Dredge, 1890.

Equinox, 'Road test', Orlando Productions for Channel 4 TV, 25 September 1988.

A.W. Evans, 'The prediction of car ownership; a comment'; *Journal of Transport Economics and Policy*: 4 (1970), 88–99.

A.W. Evans and G. Crampton, 'Myth, reality and employment in London';

Journal of Transport Economics and Policy: 23 (1989), 89–108.
M.H. Fairhurst and T. Mack, *Consequences and Causes of Changes in Road Traffic Levels. Economic Research Report R243*; London, London Transport Executive, 1981.
M.H. Fairhurst and R.S. Smith, *Development and calibration of London Transport's SCENARIO Model. Economic Research Report R229*; London, London Transport Executive, 1977.
G.J.W.A. Ferguson, 'A Comparison of Journey Times by Car and Bus in Central London—1983 Resurvey with Pilot Survey of Times by Rail'; *Traffic Engineering and Control,* 26 (1985), 198–204.
A. Fradat, J.R. Fradin, C. Cottour and Y. Farhi, *Les transports de voyageurs en Ile-de-France*; Paris, Préfecture de la région d'Ile-de-France, 1984, p.23.
M. Frank and P. Wolfe, 'An algorithm for quadratic programming'; *Naval Research Logistics Quarterly*: 3 (1956), 95–110.
T.W. Freeman, *The Conurbations of Great Britain*; Manchester, Manchester University Press, 2nd edition 1966.
A.F. Friedlaender, 'Price distortions and second-best investment rules in the transportation industries'; *American Economic Review (Papers and Proceedings)*: 71 (1981), 389–93.
S. Glaister, *Urban Public Transport Subsidies; an Economic Assessment of Value for Money. Summary report and technical report*; London, Department of Transport, 1982.
T.F. Golob, *Continuous Urban Modelling: A Critical Review*; White Paper on research in new methods of urban and regional modelling. Washington D.C., US Department of Transportation, Research and Special Programs Administration, 1982.
P.B. Goodwin, 'Car and Bus Journeys to and from Central London in Peak Hours'; *Traffic Engineering and Control*: 11 (1969), 376–78.
P.B. Goodwin and P.M. Jones, *Road Pricing; the Political and Strategic Possibilities. Systems of Infrastructure Cost Coverage (Road Pricing, Principles and Application). Round Table 80*; Paris, European Conference of Ministers of Transport, 1989.
P.B. Goodwin, L. Martin and M. Talbot, *Public Transport Modelling. Public Transport Division Report*; London, Department of Planning and Transportation/Greater London Council, 1977 (Unpublished).
P.G.G. Gray, *Private Motoring in England and Wales. Social Survey report SS 327*; London, HMSO, 1969).
Greater London Council, *The Greater London Development Plan. Report of Studies*; London, Greater London Council, 1968.
Greater London Council, *The Greater London Development Plan. Statement*; London, Greater London Council, 1969a.
Greater London Council, *Movement in London: Transport Research Studies and their Context*; London, Greater London Council, 1969b.
Greater London Council, *London: the Future and You*; London, Greater London Council, 1973.
Greater London Council, *A Study of Supplementary Licensing*; London,

Greater London Council, 1974.
Greater London Council, *Bus reliability. Report of a Joint Study by Officers of the GLC and London Transport Executive*; Greater London Council, 1976.
Greater London Council, *Transport in London—a Longer-range View. Joint Report of the Transport Committee and the Policy Resources Committee. Minutes of the Council, 29 March 1977*; London, Greater London Council, 1977. See also: Prestwood Smith and May (1977).
Greater London Council, *Traffic Monitoring Review*; annual, 1980–84.
Greater London Council, *Area Control: A Scheme for Reducing Car Traffic through Central London to Limit Congestion and Improve the Environment*; London, Greater London Council, 1983.
Greater London Council, *Public Transport in London: the Next 10 Years*; London, Greater London Council, 1985.
P. Hall, *The Development of Communications. In: Greater London*; edited by J.T. Coppock and H.C. Prince, pp.52–79, London, Faber and Faber, 1964.
B. Harrison, 'Electronic Road Pricing in Hong Kong: 3) Estimating and Evaluating the Effects'; *Traffic Engineering and Control*: 27 (1986), 13–18.
W.J. Harrison, *Central London Traffic Survey (RRL Method) April/May 1968. Research Memorandum 124*; London, Department of Highways and Transportation/Greater London Council, 1969 (Unpublished).
W.J. Harrison, C. Pell, P.M. Jones and M. Ashton, 'Some advances in model design developed for the practical assessment of road pricing in Hong Kong'; in *Transportation Research*: 20A (1986), 135-43.
P. Hathaway, *GLTS Model Calibration and Testing of Forecasts. Transport Studies Note 25*; London, Department of Planning and Transportation/Greater London Council, 1975 (Unpublished). See also: Hathaway and Quibell (1975).
P. Hathaway and R. Quibell, *GLTS Model Calibrations. Transport Studies Note 29*; London, Department of Planning and Transportation/Greater London Council, 1975 (Unpublished).
R.A. Heinlein, *The Rolling Stones*; UK and Commonwealth edition (not Canada), New York, Charles Scribner's Sons, 1952.
R.A. Heinlein, *Space Family Stone*; London, Gollancz, 1952.
B. Heydecker and G.W.J.A. Ferguson, *A Comparison of Journey Times by Car, Bus and Rail in Central London—1988 resurvey. Transport Studies Group Note*; London, University College London, 1990 (Unpublished).
D.J. Holden, 'Wardrop's Third Principle: Urban Traffic Congestion and Traffic Policy'; *Journal of Transport Economics and Policy*: 23 (1989), 239–62.
E.M. Holroyd, *Changes in Population Density in England and Wales. Laboratory Report LR1076*; Crowthorne, Berks, Transport and Road Research Laboratory, 1984.
E.M. Holroyd and D.A. Scraggs, 'Journey Times by Car and Bus in Central London'; *Traffic Engineering and Control*: 6 (1964), 169-73.

E.M. Holroyd and J.C. Tanner, *A Simplified Form of the CRISTAL Transport Planning Model. Supplementary Report SR 55UC*; Crowthorne, Berks, Transport and Road Research Laboratory, 1974.

H. Hotelling, 'Edgeworth's Taxation Paradox and the Nature of Supply and Demand Functions'; *Journal of Political Economy*: 40(5) (1932), 577–616.

H. Hotelling, 'The General Welfare in Relation to Problems of Taxation and of Railway and Utility Rates'; *Econometrica*: 6 (1938), 242–269.

House of Commons, *Minutes of Evidence Given Before the Select Committee on Metropolitan Communications*; QQ 1345–8, App. pp.215–6, Tables II and IV. London, Houses of Parliament Papers, 1855.

House of Commons, *Minutes of Evidence Given Before the Select Committee on Metropolitan Communications*; vol viii. QQ 1226–8, London, Houses of Parliament Papers, 1863.

House of Commons, *Minutes of Evidence Given Before the Select Committee on the Post Office (London) Railway Bill, 3rd July 1913. Report and Special Report from the Select Committee etc. together with the Proceedings and Minutes of Evidence*; London, Houses of Parliament Papers, Wyman for HMSO, 1913.

J. Inwood, *Some Effects of Increased Parking Meter Charges in London 1965. Road Research Laboratory Report LR* 7, Crowthorne, Berks, now Transport and Road Research Laboratory, 1966.

R.H. Johnston, 'Some Mechanisms of Speed Similarity in Urban Areas; 1) The Similarity of Observed "Door-to-Door" Speeds: 2) The Stability of Observed Road Speeds'; *Traffic Engineering and Control*: 29 (1988), 1) 6–9: 2) 72–75.

S. Joy, *The Train that Ran Away; a Business History of British Railways 1948–1968*; London, Ian Allan, 1973.

S. Joy, 'Railway Costs and Planning'; *Journal of Transport Economics and Policy*: 23 (1989), 45–54.

E. Kant, 'Zur Frage der inneren Gliederung der Stadt, insbesondere der Abgrenzung des Stadtkerns mit Hilfe der bevölkerungskartographischen Methoden'; *Lund Studies in Geography*, Series B, Human Geography No 24: Proceedings of the IGU Symposium in Urban Geography, Lund, 1960, edited by K. Norburg, pp.321–81. Gleerup for Department of Geography, Royal University of Lund, 1962.

J.R. Kellett, *Railways and Victorian Cities*; paperback edition, London, Routledge and Kegan Paul, 1979.

G. Kemmann, *Der Verkehr Londons mit besonderer Bercksichtigung der Eisenbahnen*; Berlin, 1890, p.180. In: P. Hall, *The Development of Communications in Greater London*; edited by J.T. Coppock and H.C. Prince, London, Faber and Faber, 1964, p.69.

F.H. Knight, 'Some fallacies in the interpretation of social cost'; *Quarterly Journal of Economics*: 38 (1924), 582–606. Reprinted in F.H. Knight, *The Ethics of Competition*; New York, Harper, 1935: and in the *American Economic Association, Readings in Price Theory*; selected by G.J. Stigler and K.E. Boulding, Chicago, Richard Irwin, 1952, pp.160–179; and

London, George Allen and Unwin, 1953.
N.D. Kondratiev, 'Die lange Wellen der Konjunktur'; *Archiv für Sozialwissenschaft und Sozialpolitik*: 56 (1926), 573–609. In English: *Review of Economic Statistics*; 17 (1935), 105–115.
O. Lambert, Letter to *The Times*, 27 April 1988.
R.J. Lane, T.J. Powell and P. Prestwood Smith, *Analytical Transportation Planning*; London, Duckworth, 1971.
A.S. Lang and R.M. Soberman, *Urban Rail Transit; its Economics and Technology*; Cambridge, Massachusetts, M.I.T. Press for the Joint Centre for Urban Studies, M.I.T. and Harvard University, 1966.
D. Lardner, *Railway economy; a Treatise on the New Art of Transport, its Management, Prospects and Relations, Commercial, Financial and Social*; London, 1850.
H. Leavitt, *Superhighway—Superhoax*; Garden City, New York, Doubleday, 1970.
D.L. Lewis, 'Public Policy and Road Traffic Levels in Greater London'; *Journal of Transport Economics and Policy*: 11 (1977), 155–68.
London Passenger Transport Research Group, *Route 31 study*; London, 1983.
London Regional Transport, *Review of Service Performance: Third Quarter 1989*; London, Group Planning Office, LRT, 1989 (Unpublished).
London Traffic Surveys 1947–66. See: Newby and Wardrop (1947), Smeed (1961), Turner and Crawford (1966), Thomson (1967a).
London Traffic Survey 1962; volume 1, chapters 1–9, London, London County Council, 1964.
London Traffic Survey 1962; volume 2, chapters 10–18, London, 1966. See: *London Transportation Study* (1968) and Greater London Council (1969b).
London Transport Executive, *London Travel Survey 1949*; London, Research Services Ltd for LTE, 1950.
London Transport Executive, *London Travel Survey 1954*; London, LTE, 1956.
London Transport Executive, *London Transport's Corporate Aim*, Group Planning; London, LTE, 1977.
London Transport Executive, *Annual Report*; London, LTE, 1982.
London Transportation Study; volume 3, chapters 19–24 (4 parts), London, Freeman, Fox, Wilbur Smith and Associates for the Greater London Council, 1968.
D.A. Lynam, *The Application of CRISTAL to Evaluate Changes in Travel Charges in London. Supplementary Report SR40UC*; Crowthorne, Berks, Transport and Road Research Laboratory, 1974.
D.A. Lynam and P.F. Everall, *Public Transport Journey Times in London—1970. Laboratory Report LR 413*; Crowthorne, Berks, Road Research Laboratory (now Transport and Road Research Laboratory), 1970a.
D.A. Lynam and P.F. Everall, *Car Journey Times in London—1970. Laboratory Report LR 416*; Crowthorne, Berks, Road Research Laboratory (now Transport and Road Research Laboratory), 1970b.

A. Marshall, *Principles of Economics*; London, Macmillan, 1890.

A.D. May and I. Turvey, *Personal Communication*; Leeds, Institute for Transport Studies, 1986.

C.S. Meik and W. Beer, *The Improvement of London Traffic*; London, Royal Society of Arts, 1898. Revised in: *The Improvement of London Traffic*; London, P.S. King, 1905. Summarised in: Buchanan (1970).

M.J.H. Mogridge, 'The Prediction of Car Ownership'; *Journal of Transport Economics and Policy*: 1 (1967), 52–74. See also: A.W. Evans (1970) and M.J.H. Mogridge (1970a).

M.J.H. Mogridge, *An Examination of some of the Factors Influencing the Spatial Income Distribution of Households in a City Region. Working Paper 7*; London, Centre for Environmental Studies, 1968. Revised version in: 'Studies in Regional Science'; edited by A.J. Scott, *London Papers in Regional Science*: 1 (1969), 117–41, London, Pion.

M.J.H. Mogridge, 'The prediction of car ownership: a rejoinder'; *Journal of Transport Economics and Policy*: 4 (1970a), 100–06.

M.J.H. Mogridge, *SCAMP—Shorter Curvilinear Alternative Minimum Paths—a Computer Algorithm. Urban Traffic Model Research, 143–145. Planning and Transport Research and Computation conference paper*; London, PTRC, 1970b.

M.J.H. Mogridge, *The Transportation Planning Process—a Critical Appraisal Using a Problem from Hong Kong as an Example (the Cross-harbour Ferry Study). Technical Report 93*; Hong Kong, Traffic and Transport Survey Unit, Public Works Department, Hong Kong Government, 1971.

M.J.H. Mogridge, 'Some Thoughts on the Intra-urban Spatial Location of Homes, Worker Residences and Workplaces'; *Conference Paper*: 9 (1973), 249–307 (First Urban Economics Conference, Keele University), London, Centre for Environmental Studies.

M.J.H. Mogridge, *Changing Spatial Patterns in the Journey-to-work; a Comparison of the 1966 and 1971 Census Data in London*: P168 (1978), 52–69, London, PTRC Education and Research Services Ltd for Planning and Transportation Research and Computation (International) Co. Ltd. Revised in: *Urban Studies*; 16 (1979), 179–90.

M.J.H. Mogridge, 1) 'The Joint Effect of Land Use and Distance from the Centre of London on Journeys by Bus and Journeys to Work by all Modes: an Analysis of 1971 GLTS Data'. 2) 'The Joint Effect of Land Use and Distance from the Centre of London on Journeys by Bus: Further Analysis of GLTS Data on Travel Times and Speeds'; *Transport Studies Group Notes*, London, University College London, 1980a.

M.J.H. Mogridge, *Analyse des Caracteristiques des Déplacements et de leur Relation à la Structure Urbaine: une Comparaison entre Paris et Londres*; Paris, Martin Mogridge Associates for Institut de Recherche des Transports, 1980b.

M.J.H. Mogridge, *Analyse des Déplacements Domicile-Travail dans l'Agglomération Parisienne à Partir de Coordonnés Polaires*; Paris, Martin Mogridge Associates for Institut de Recherche des Transports, 1980c.

M.J.H. Mogridge, *The Car Market; a Study of the Statics and Dynamics of*

Supply/Demand Equilibrium; London, Pion, 1983a.

M.J.H. Mogridge, *Review of Work Comparing Transport and Travel in Paris and London. Transport Studies Group Note*; London, University College London for Transport and Road Research Laboratory, 1983b (Unpublished).

M.J.H. Mogridge, *Review of Work Comparing Transport and Travel in Paris and London: Supplementary Comments on Paris and London. Transport Studies Group Note*; London, University College London for Transport and Road Research Laboratory, 1984 (Unpublished).

M.J.H. Mogridge, 'Strategic Population Forecasting for a Conurbation using the Negative Exponential Density Model'; *Transportation Research*: 19A (1985), 189–206.

M.J.H. Mogridge, 'Jam yesterday, jam today and jam tomorrow? (or how to improve traffic speeds in Central London)'; *University College London Lunch-Hour Lecture, 17 October 1985*, London, Martin Mogridge Associates for Transport Studies Group, University College London, 1986a.

M.J.H. Mogridge, 'If London Is More Spread Out Than Paris, Why Don't Londoners Travel More than Parisians?'; *Transportation*: 13 (1986b) 85–104.

M.J.H. Mogridge, *Linking the BR Network Across Central London. Transport Studies Group Note*; London, University College London, 1988a (Unpublished).

M.J.H. Mogridge, *The Variability of Rail Journey Times on Particular Routes to Central London and the Implications for Modal Split Equilibrium Processes. Transport Studies Group Note*; paper presented at the Universities Transport Studies Group Annual Conference 1988; London, University College London, 1988b (Unpublished).

M.J.H. Mogridge, 'The Prediction of Car Ownership and Use Revisited: the Beginning of the End?'; *Journal of Transport and Policy*: 23 (1989), 55–74.

M.J.H. Mogridge and J.J. Bates, *Recalibration of the 1962 London Traffic Survey*; London, Martin Mogridge Associates and John Bates Services for Department of Transport, 1980.

M.J.H. Mogridge and D. Eldridge, *Car Ownership in London. Research Report 10*; London, Intelligence Unit/Greater London Council, 1970.

M.J.H. Mogridge and S.A.W. Fry, 'The Variability of Car Journey Times on a Particular Route in Central London'; *Traffic Engineering and Control*: 25 (1984), 510–11.

M.J.H. Mogridge and D.J. Holden, *Spatial Variations in Average Daily Travel Times in London. Transport Studies Group Note*; paper presented at the Universities Transport Study Group Annual Conference 1983, London, University College London, 1983 (Unpublished).

M.J.H. Mogridge and D.J. Holden, 'A Panacea for Road Congestion: a Riposte'; *Traffic Engineering and Control*: 28 (1987), 13–19.

M.J.H. Mogridge, D.J. Holden, J. Bird and G.C. Terzis, 'The Downs-Thomson Paradox and the Transportation Planning Process'; *Inter-*

national Journal of Transport Economics: 14 (1987), 283–311.

H. Mohring, 'Optimization and Scale Economies in Urban Bus Transportation', *American Economic Review*: 62 (1972), 591-604.

Movement for London, *Wheels for Change*; London, British Road Federation, 1987.

P.W. Munt and R. Woodhall, *A Study of Travel Times by Private and Public Transport in Greater London. Research Memorandum 44*; London, Department of Highways and Transportation/Greater London Council, 1967.

MVA Consultancy, The, *A Technical Audit of the GLTS Model—Final Report*; London, MVA Systematica for the Department of Transport, 1988.

MVA Consultancy, The, *Summary of LTS Model Forecasts. London Assessment Studies Stage 2B. LTS Technical Note 114*; London, The MVA Consultancy for the Department of Transport, 1989.

MVA Consultancy, The, *Transportation Strategic Advice: Scenario Testing Exercise. 3) Evaluation of Options Using the LTS Model*; London, The MVA Consultancy for the London Planning Advisory Committee, 1990.

MVA Consultancy, The, Colin Buchanan and Partners and the Transport and Road Research Laboratory, *Transportation Strategic Advice: Scenario Testing Exercise. 1) Evaluation of Options Using the London Area Model*; London, The MVA Consultancy, Colin Buchanan and Partners and the Transport and Road Research Laboratory for the London Planning Advisory Committee, 1988.

MVA Consultancy, The and Colin Buchanan and Partners, Transportation Strategic Advice: Scenario Testing Exercise. 2)Desk studies of transportation topics; London, The MVA Consultancy and Colin Buchanan and Partners for the London Planning Advisory Committee, 1989.

National Travel Surveys (1964/66, 1972/73, 1975/76, 1978/79, 1984/85); London, HMSO.

R.F. Newby, *A Note on the Distribution of Traffic Speeds. Research Note*; Crowthorne, Berks, Road Research Laboratory (now Transport and Road Research Laboratory), 1949 (Unpublished).

R.F. Newby, 'Road traffic in Central London'; *PIARC Journal*: 43(136) (1954), 1–18.

R.F. Newby and J.G. Wardrop, *London Traffic Survey 1947. Research Note*; Crowthorne, Berks, Road Research Laboratory (now Transport and Road Research Laboratory), 1947 (Unpublished). Later surveys follow in the same research note series.

B. Newling, 'Urban Population Densities and Intra-urban Growth'; *Geographical Review*: 54 (1964), 440–42.

B. Newling, 'Urban Growth and Spatial Structure: Mathematical Models and Empirical Evidence'; *Geographical Review*: 56 (1966), 213–25.

B. Newling, 'The spatial variation of urban population densities'; *Geographical Review*, 59 (1969), 242–52.

J.P. Nicholl, 'Journey Times by Car and Bus in Central London—1977 Resurvey'; *Traffic Engineering and Control*: 18 (1977), 581-83.

R.H. Oldfield, P.H. Bly and F.V. Webster, *Predicting the Use of Stage Service Buses in Britain. Laboratory Report LR 1000*; Crowthorne, Berks, Transport and Road Research Laboratory, 1980.

P. Olszewski and W. Suchorzewski, 'Traffic Capacity of the City Centre'; *Traffic Engineering and Control*: 28 (1987), 336–37, 339–43, 348. See also Letters, 647.

V. Outram, *An Analysis of the Spatial Distribution of the 1966 Census Population Data of the Towns of Great Britain using Polar Coordinates. Systems Analysis Research Unit Note*; London, Department of the Environment, 1972 (Unpublished).

W. Owen, *The Metropolitan Transportation Problem*; Washington D.C., Brookings Institution, 1956 (see for example p.253).

W. Owen, *The Accessible City*; Washington D.C., Brookings Institution, 1972, pp.15–21.

C.N. Parkinson, *The Law and the Profits*; Boston, Massachusetts, Houghton, 1960.

J.B. Parr, 'A Population Density Approach to Regional Spatial Structure'; *Urban Studies*: 22 (1985), 289–303.

Pathé News, 'The Great Hold-up' (Newsreel 53/98), 1953. In: *Equinox* 'Fatal Attraction'; Juniper Productions for Channel 4 T.V., 13 August 1989.

A.C. Pigou, *The Economics of Welfare*; London, Macmillan, 1920 (see especially p.194).

Postmaster-General, *Report of the Committee Appointed by the Postmaster-General to Consider the Question of the Transmission of Mail in London by Pneumatic Tubes or Underground Electric Railways. Min. No. 22, 242/09, 446, 096/09*; London, General Post Office, 1911.

A.R. Prest and R. Turvey, 'Cost-benefit Analysis; a Survey'; *Economic Journal*: 75 (1965), 683–735.

P. Prestwood Smith, 'Area control'; *Traffic and Environmental Management*: P180 (1979), 37–49, London, PTRC Education and Research Services Ltd for Planning and Transportation Research and Computation (International) Co. Ltd.

P. Prestwood Smith and A.D. May, 'The Assessment of Transport Programme Options'; *Transport Planning Practice*: P155 (1977), 43–60, London, PTRC Education and Research Services Ltd for Planning and Transportation Research and Computation (International) Co. Ltd.

D.A. Quarmby, 'Choice of Travel Mode for the Journey to Work; Some Findings'; *Journal of Transport Economics and Policy*: 1 (1967), 273–314.

A.U. Reid and J.G. Cottee, *A Procedure for the Design of a Road Network. Report HT/T-1*; London, Traffic Branch, Department of Highways and Transportation/Greater London Council, 1967.

G. Rendle, T. Mack and M.H. Fairhurst, *Bus and Underground Travel in London; an Analysis of the Years 1966–1976*; Economic Research Report R235, London LTE, 1978.

P. Rice, 'Queueing and Stochastic Aspects of Urban Railway Capacity with

Special Reference to London Underground Railways (unpublished Ph.D. thesis, University College London, 1974).

W.Z. Ripley, *Railroads: Rates and Regulations*; (reprint), New York, Longmans Green, 1916, pp.55–56.

G. Roth, *The Economic Benefits to be Obtained by Road Improvements with Special Reference to Vehicle Operating Costs. Report to the Rees Jeffreys Road Fund. Research Note*; Crowthorne, Berks, Road Research Laboratory, Department of Scientific and Industrial Research (now the Transport and Road Research Laboratory), 1959 (Unpublished.)

Royal Commission on Metropolis Railway Termini, *Report*; London, HMSO, 1846.

Royal Commission on London Traffic, *Report*; London, HMSO, 1906.

M.L. Senior and H.C.W.L. Williams, 'Model-based Transport Policy Assessment; 1) The Use of Alternative Forecasting Models: 2) Removing Fundamental Inconsistencies from the Models'; *Traffic Engineering and Control*: 18 (1977), 1) 402–06: 2) 464–69. See also: Williams (1977).

C.H. Sharp, 'The Choice Between Cars and Buses on Urban Roads'; *Journal of Transport Economics and Policy*: 1 (1967), 104–11.

R.J. Smeed, *The Traffic Problem in Towns. Norbury, Lockwood, Cheetham, Manchester*; for Manchester Statistical Society. Revised as: 'The Traffic Problem in Towns—A Review of Possible Long Term Solutions'; *Town Planning Review*: 35 (1964), 133–58.

R.J. Smeed, *Road Pricing; The Economic and Technical Possibilities*; London, HMSO, 1964).

R.J. Smeed, 'Traffic Studies and Urban Congestion'; (inaugural address as Professor of Traffic Studies, University College London), *Journal of Transport Economics and Policy*, 2 (1968), 33–70.

R.J. Smeed and G.O. Jeffcoate, 'The Variability of Car Journey Times on a Particular Route'; *Traffic Engineering and Control*: 13 (1971), 238-43.

R.J. Smeed and J.G. Wardrop, 'An Exploratory Comparison of the Advantages of Cars and Buses for Travel in Urban Areas'; *Journal of the Institute of Transport*: 30(9) (1964), 301–315. And in: *Revue de la Fédération Internationale de l'Automobile*, 1 (1965); and *Automobilismo*, 13(1) (1965), new series. 62–98.

A.A. Stroud, *Initial Processing of GLTS Household Interview Data. Research Memorandum 310*; London, Department of Planning and Transportation/ Greater London Council, 1974.

W. Suchorzewski, 'Principles and Applicability of the Integrated Transportation System'; *Proceedings of the UN-ECE Seminar on the Role of Transportation in Urban Planning, Development and Environment*, Muenchen, 1973.

W. Suchorzewski, 'Planning Measures to Promote a Balance Among Traffic and Transportation Modes with Special Regard to the Human Scale'; *Proceedings of the 2nd UN-ECE Seminar on the Role of Transportation in Urban Planning, Development and Environment*, Washington D.C., 1976.

J.C. Tanner, L. Gyenes, D.A. Lynam, S.V. Magee and A.H. Tulpule,

Development and Calibration of the CRISTAL Transport Planning Model. Laboratory Report LR574, Crowthorne, Berks, Transport and Road Research Laboratory, 1973. See also: Holroyd and Tanner (1974), Lynam (1974).

G.C. Terzis and M.J.H. Mogridge, *The Use of Panel Surveys to Study the Variability of Travellers' Behaviour. Transport Studies Group Note*; (paper presented at the Universities Transport Studies Group Annual Conference), London, University College London, 1988 (Unpublished).

W.S. Thompson, *Population, the Growth of Metropolitan Districts in the United States, 1900–1940*; Washington D.C., U.S. Government Printing Office, 1947.

J.M. Thomson, 'Speeds and Flows of Traffic in Central London: 1) Sunday Traffic Survey; 2) Speed-flow Relations'; *Traffic Engineering and Control*: 1) 8(11) (1967a), 672–76; 2) 8(12) (1967a), 721–25.

J.M. Thomson, 'An Evaluation of Two Proposals for Traffic Restraint in Central London (with Discussion)'; *Journal of the Royal Statistical Society*: 130A (1967b), 327–77.

J.M. Thomson, *Motorways in London. Report of the Working Party Led by J.M. Thomson*; London, Duckworth for London Amenity and Transport Association, 1969.

J.M. Thomson, *Great Cities and their Traffic*; London, Gollancz, 1977, Peregrine edition, p.165.

'Traffic Engineering and Control News'; *Traffic Engineering and Control*: 20 (1979), 366.

K. Train, 'Optimal Transit Prices under Increasing Returns to Scale and a Loss Constraint'; *Journal of Transport Economics and Policy*: 11 (1977), 185–94.

Transport and Road Research Laboratory, *The Demand for Public Transport; Report of the International Collaborative Study of the Factors Affecting Public Transport Patronage*; edited by F.V. Webster and P.H. Bly, Crowthorne, Berks, Transport and Road Research Laboratory, 1980.

Transport and Road Research Laboratory, *Cost/Benefit Assessment of Bus Service Subsidy. Working Paper*; Crowthorne, Berks, Special Research Branch/Transport and Road Research Laboratory, 1983 (Unpublished).

Transport and Road Research Laboratory. See: The MVA Consultancy (1988).

E.D. Turner and K. Crawford, *Central London Traffic Survey, April-May 1966. Research and Development Paper 4.* London, Department of Highways and Transportation/Greater London Council, 1966.

D. Van Vliet and P.D.C. Dow, 'Capacity-restrained Road Assignment'; *Traffic Engineering and Control*: 20 (1979), 296–305.

R.J. Vaughan, *Urban Spatial Traffic Patterns*; London, Pion, 1987.

W.S. Vickery, 'Congestion Theory and Transport Investment'; *American Economic Review*: 59 (1969), 251–60.

W.S. Vickery, 'Optimisation of traffic and facilities', *Journal of Transport Economics and Policy*, 1 (1967), 123–36.

A.A. Walters, 'Track Costs and Motor Taxation'; *Journal of Industrial Economics*, 1954.

J.G. Wardrop, 'Some Theoretical Aspects of Road Traffic Research'; *Proceedings of the Institution of Civil Engineers (ICE), Road Paper*: 36, with discussion (1952). 325–78, London, I.C.E.

J.G. Wardrop, 'Journey Speed and Flow in Central Urban Areas'; *Traffic Engineering and Control*: 9 (1968), 528–32, 539.

J.G. Wardrop and J. Turner, '1) The Variation of Journey Time in Central London. 2) The Variation of Journey Speed in Central London'; *Research Notes*, Crowthorne, Berks, Road Research Laboratory (now the Transport and Road Research Laboratory) 1951 (Unpublished).

T.M. Watson, W.S. Smith and F.W. Hurd, *Traffic Engineering*; New York, McGraw-Hill, 1955.

A.F. Weber, *Growth of Cities in the Nineteenth Century*; London, Macmillan, 1899. New York; reprinted by Cornell University Press, Ithaca, New York, 1963.

F.V. Webster, 'Traffic Signal Settings'; *Technical Paper 39*, Crowthorne, Berks, Road Research Laboratory/Department of Scientific and Industrial Research (now Transport and Road Research Laboratory). Reprinted with minor amendments (1961). Updated as: F.V. Webster and B.M. Cobbe, 'Traffic Signals'; *Technical Paper 56*, 1966.

F.V. Webster and R.H. Oldfield, 'A Theoretical Study of Bus and Car Travel in Central London'; *Transport and Road Research Laboratory Report LR 541*, Crowthorne, Berks, Transport and Road Research Laboratory, 1972.

H.K. Weiss, 'The Distribution of Urban Population and an Application to a Servicing Problem'; *Operations Research* 9 (1961), 860–74.

A.M. Wellington, *Great Economic Theory of the Location of Railways*; New York, Wiley, 1889.

W.C. Wheaton, 'Price-induced Distortions in Urban Highway Investment'; *Bell Journal of Economics*, 9 (1978), 622–32.

H.C.W.L. Williams, 'On the Formation of Travel Demand Models and Economic Evaluation Measures of User Benefit'; *Environment and Planning A*, 9 (1977), 285–344.

Wootton Jeffreys Consultants Ltd., 'A Study to Show Patterns of Vehicle Use in the London Area'; *Contractor Report CR131*, supplement CR132, Crowthorne, Berks, Transport and Road Research Laboratory.

L.B. Yates and T.J. Howerd, 'Journey Times by Car in Central London'; *Research Memorandum 150*, London, Department of Highways and Transportation/Greater London Council, 1969 (Unpublished).

Index